Dangerous Miracle

Dangerous Miracle

A Natural History of Antibiotics – and how we burned through them

LIAM SHAW

THE BODLEY HEAD
LONDON

1 3 5 7 9 10 8 6 4 2

The Bodley Head, an imprint of Vintage, is part of the
Penguin Random House group of companies

Vintage, Penguin Random House UK, One Embassy Gardens,
8 Viaduct Gardens, London SW11 7BW

penguin.co.uk/vintage
global.penguinrandomhouse.com

First published by The Bodley Head in 2025

Typeset in 12/14.75pt Dante MT Std by Six Red Marbles UK, Thetford, Norfolk
Printed and bound in India by Manipal Technologies Limited

The authorised representative in the EEA is Penguin Random House Ireland,
Morrison Chambers, 32 Nassau Street, Dublin D02 YH68

A CIP catalogue record for this book is available from the British Library

HB ISBN 9781847927545
TPB ISBN 9781847927552

Penguin Random House is committed to a sustainable future
for our business, our readers and our planet. This book is made
from Forest Stewardship Council® certified paper.

To my parents

Table of contents

Let us not, however, flatter ourselves overmuch on account of our human conquest of nature. For each such conquest takes revenge on us.

Friedrich Engels

Introduction

Fossil drugs

The Klondike is a region of north-west Canada where the winters are long and cold. If you dig six feet into the ground you will hit the permafrost: frozen soil under the soft surface, iron-hard any day of the year. In the late nineteenth century, prospectors flooded to the region, drawn by news of gold in the ground, but they couldn't dig past the permafrost. Instead, they dug holes in the topsoil, then set fires and watched them burn. Through the long winter nights, flames raged across the valleys as the fires melted down through the earth itself.

Without knowing it, the prospectors were melting history. Permafrost is a record of the past. Its layers hold ancient DNA in ideal preservation: the deeper you go into the ground, the further back you go in time, like rifling through a frozen stack of Ice Age postcards.

On a summer day in 2007, just a few kilometres from where the Klondike Gold Rush had begun, a group of scientists stood in front of a cliff of permafrost. It had been exposed by twentieth-century mining, when miners had moved on from shovels to more powerful machines that split open the frozen earth. The scientists approached the site like forensic investigators at a crime scene. They drilled horizontally into the cliff, pulling out a short cylinder of 30,000-year-old permafrost. After shipping it back to McMaster University in Ontario, they worked on it in a dedicated cleanroom. They carefully sterilised their tools, then cut two small incisions across the icy cylinder and split it using a hammer and chisel: out popped a disc as thick as a hamburger. They punched chunks out of the hamburger and dissolved them in hot chemicals. At that point, frozen cells in the soil that had been locked away for thousands of years melted, burst and released their DNA.

When the scientists sequenced the DNA, a whole world shimmered out of the fragments, summoning the molecular ghosts of extinct bison and woolly mammoths that had once wandered through a lost landscape of sage-grass and willow. But the scientists hadn't gone in search of the ghosts of mammoths. They were looking for antibiotic resistance.

From the beginning, we have shared our world with trillions of microbes. Among them are the invisible assailants responsible for infectious diseases, which range from trivial ailments to the most notorious scourges of humanity. Some are viruses such as HIV: tiny bundles of genetic material with no cells of their own, which hijack our cellular machinery to churn out copies of themselves. Some are fungi, such as the *Trichophyton* moulds that cause athlete's foot. Some are parasites, such as the *Plasmodium* species that cause malaria. And some are bacteria.

Bacteria live almost everywhere. They can be found on land, from the deepest gold mines to the highest peaks; in air, from the clouds over Antarctica to a packed commuter train; in water, from near-boiling geysers to the ocean's darkest abyss. They can also be found inside us, coating our gums and lining our guts. There are far more bacterial species than science has reliably identified, but we know that most of the tens of thousands we have named are harmless. However, a minority have the capacity to cause infection and disease. In contrast to the blitzkrieg of deaths caused by a viral pandemic such as Covid-19, bacterial infections are a slow attritional war. The number of culprits is small, but their impact is colossal: just over thirty bacterial species are linked to 1 in 8 of all deaths worldwide. Invisible to the naked eye, these minuscule creatures with strange and unwieldy names – *Staphylococcus aureus*, *Escherichia coli*, *Streptococcus pneumoniae* – are some of the world's biggest killers. Our best defences against them are antibiotics.

An antibiotic is any medicine that works to kill or incapacitate bacteria.* Starting in the 1930s, effective antibiotics caused a transformation in human history without precedent. Within just a few decades, the world's worst infectious diseases could be cured. Writing in the aftermath of that revolution, one British doctor in 1968 commented that 'no one recently qualified, even with the liveliest imagination, can picture the ravages of bacterial infection which continued until little more than thirty years ago'. Antibiotics did more than vanquish diseases; they permitted the entire edifice of modern medicine to be built up on their foundation.

Consider surgery. Cutting people open and breaking the protective barrier of the skin had always been astonishingly dangerous, because it gave bacteria the chance to swarm into the body's internal tissues. Before antibiotics, even the simplest procedures could result in blood poisoning, sepsis and death. With antibiotics, surgery became significantly safer and its remit expanded: heart surgery, intestinal surgery and liver transplants became routine. Or consider cancer: chemotherapy suppresses the immune system, leaving patients vulnerable and exposed to bacterial attack. Infections are one of the biggest complications of cancer care, and one in five cancer patients will receive antibiotics during their treatment. Cancer and surgery are not exceptions. In almost every area of medicine, the sophistication of our most advanced treatments relies on antibiotics.

However, not every antibiotic kills every bacterium. The diversity of different bacterial species is reflected by the diversity of different antibiotics, resembling the range of weapons available in an armoury. There are over sixty different antibiotics in the World Health Organization's list of essential medicines, each effective against a slightly different range of bacteria. A bacterium that can survive treatment with an antibiotic is *resistant* to it. Bacteria can

---

* Some scientists use 'antibiotic' to refer only to natural products produced by living organisms; for a more general term for bacteria-killing substances, they prefer 'antibacterial' or 'antimicrobial'. In this book, I follow common usage and call all these medicines antibiotics.

either be resistant intrinsically – just as rock is ever impervious to scissors – or evolve to become so. They can resist an antibiotic in a number of ways: they can modify what it attacks, destroy it directly, pump it out of the cell before it can do damage, or stop it getting inside in the first place. The details of any particular resistance can be intricate, but the outcome is simple: antibiotic treatment fails.

From the introduction of antibiotics, doctors were horrified to find that this was possible. It seemed as if antibiotics themselves were capable of conjuring up even more fearsome bacteria. But resistance is nothing more than a simple consequence of evolution. Each time an antibiotic was used, it killed off the bacteria that could be killed and left behind those that could not. Whether because of a chance mutation in their genes or the possession of different genes altogether, they remained unaffected. Using an antibiotic caused susceptible bacteria to diminish and resistant bacteria to thrive, and those resistant bacteria could spread between people. Tragically, the miraculous powers of these new medicines could wane over time.

Over the twentieth century, each new antibiotic seemed to give rise to new resistance genes. Where were they coming from? In search of the origins of resistance, scientists turned to the freezers that lurked in the corners of universities, hospitals and departments of public health. These freezers contained old samples of bacteria from the start of the twentieth century, from before the discovery of antibiotics: samples from Italian dysentery epidemics or the ghastly wounds of WWI soldiers. When scientists searched these archives, they found that the bacteria in the freezers had few if any of the genes that caused modern antibiotic resistance. It was as if the introduction of antibiotics had prompted sudden changes in bacteria. Antibiotic resistance appeared to be a thoroughly modern story of technological hubris.

Others disagreed. In 2006, the Canadian microbiologist Gerry Wright showed that handfuls of dirt collected far from human settlements contained antibiotic-resistant bacteria in abundance. To many, that was an alarming surprise. But a common criticism was that no corner of the world had been unaffected by humans: how could Wright be sure that the soil hadn't somehow been contaminated with antibiotics? To address that objection, Wright needed to analyse soil from the deep past.

That was where the Klondike came in. Wright and his colleagues hoped to use the permafrost as a natural freezer. When they analysed its ancient frozen DNA, they found something extraordinary. Amid the fragments of mammoth and bison there were bacteria that were resistant to modern antibiotics. The scientists even found resistance genes against an antibiotic called vancomycin, which was relied upon as a last-ditch treatment for the resistant superbug known as MRSA. Vancomycin had first been used in 1955 – but the permafrost did not lie.

It seemed like a paradox. How could frozen bacteria from 30,000 years ago be resistant to an antibiotic created in the 1950s? The answer was that antibiotics were not invented in the twentieth century. They were far, far older.

Humans discovered antibiotics in the twentieth century, but we did not invent them. Most of the important antibiotics we rely on today are based on molecules made by single-celled microbes – the same forms of life that cause infectious disease. Killing one of these cells is easy: you can boil it into oblivion or dissolve it with acid. But we cannot use boiling water or acid to kill bacteria inside the body, because that would also harm us. This is where antibiotics are different. Over millions of years of evolution, bacteria and fungi evolved molecules to kill, maim or incapacitate each other. But our last common ancestor with bacteria lived billions of years ago. Evolution since has made the basic cellular architecture of humans very different from bacteria. A human cell has evolved to live only within the fluid maintained inside human bodies, where the concentration of molecules outside the cell balances what's inside. If you place a cell from your body into pure water, it will slowly expand as water rushes in through its permeable membrane, before bursting like an overfilled balloon. In contrast, bacteria have a cell wall which keeps them rigid in water, allowing them to survive in many different environments.

Antibiotics evolved in nature to target cellular components that bacteria possess and we do not. That simple fact means we can safely ingest antibiotics and leave our own cells largely unscathed. In this way, antibiotics are unique among medicines. Most drugs do their work by interfering with human cells: aspirin relieves your headache by dampening the chemical signals of pain and inflammation; caffeine wakes you up by blocking adenosine receptors, preventing drowsiness taking hold. In contrast, an antibiotic targets bacteria – an ancient curse in a language that humans no longer speak.

Antibiotics may be ancient molecules, but their deployment by humans is modern. Although history contains examples of people using antibiotic-producing moulds to treat wounds, the twentieth century was when humanity discovered how to concentrate antibiotics into medicine. That advance marks a division of history into two eras: before and after antibiotics. Like most scientific advances, it did not happen in one fell swoop. But if you had to pick a single year in which to draw the line, a reasonable choice would be 1928, when Alexander Fleming noted the activity of a strange mould juice that he called penicillin. As scientists began looking in earnest for antibiotics in the following decades, they discovered that soil contained an unimaginable bounty. By the middle of the twentieth century, what historians call the 'golden age' of antibiotics was in full swing.

The century since Fleming's observation has been the antibiotic era – what historian Claas Kirchhelle has dubbed the Antibiocene. It is a young era: if Earth's history was compressed into a single year, penicillin was discovered well into the final second. But antibiotics did not eliminate bacteria entirely. The biologist Joshua Lederberg once described microbes as 'our competitors of last resort for domination of the planet' – and the competition is far from over. Increasingly, there are warnings that rising levels of antibiotic resistance mean we may be approaching a third era: the post-antibiotic age.

The challenge of antibiotic resistance has often been

compared to climate change in its scale and scope. Indeed, there are striking similarities between fossil fuels and antibiotics. Both are natural resources that developed over millions of years before being extracted by humans. Both were mined heavily, fossil fuels powering the global economy and antibiotics fuelling modern medicine. And both offered a promise of cheap, miraculous and never-ending power that was taken as evidence of human ingenuity and mastery over the natural world. But that, of course, is only half the story. For both fossil fuels and antibiotics, their extraordinary power emerges not from human technology, but from their origins in life's distant past. As the Earth's geology transmuted dead plants into dense fuel, so evolution transmuted dead atoms into living molecules that we could use as medicines. Deep time – the near-unimaginable eons of life on Earth before humans – is a hidden wellspring from which modern civilisation flows. For the past hundred years, the extraordinary accomplishments of humanity have been powered by burning through the vastness of the past. We have been living, quite literally, on borrowed time. Antibiotics are the fossil fuels of medicine: they are fossil drugs.

To treat an antibiotic-resistant infection you need another antibiotic. Given the threat of rising antibiotic resistance, you might imagine that the discovery of new antibiotics would be proceeding at pace. The opposite is true. The dates paint a concerning picture: over half of essential antibiotics were discovered more than sixty years ago. Only one in six was discovered in the last thirty years.

Antibiotics are no longer being developed in anything like the numbers that scientists and doctors believe is necessary. In 2021, one survey of new medicines in development counted more than 4,000 molecules for cancer immunotherapy compared to fewer than forty antibiotics in clinical trials. Worse, most of those 'new' antibiotics were derivatives of existing medicines, meaning they would probably be vulnerable to similar forms of antibiotic resistance. In 2024, a report concluded that the number of scientists carrying out the crucial early stages of research into new antibiotics was 'tiny'. The world is running out of antibiotics, and new ones aren't arriving quickly enough. Understanding how and why this happened is one of the themes of this book.

As the number of effective antibiotics diminishes, we face a corresponding rise in the number of deaths from bacterial infection. In 2024, the United Nations General Assembly agreed a declaration that called antibiotic resistance 'one of the most urgent global health threats'. According to our best estimates, resistance is directly responsible for at least a million deaths a year worldwide. Over the next twenty-five years, drug-resistant infections will kill 40 million people, double the number that died in WWI. By 2050, forecasts predict that over 8 million people a year will die after contracting a resistant infection, and a quarter of those deaths will be for want of an effective antibiotic. Without new antibiotics, the entire structure of modern medicine will begin to crumble.

Recent patient stories from around the world reveal how antibiotic resistance can affect anyone, anywhere. Vanessa Carter from South Africa survived a serious car crash and needed a decade of facial reconstruction surgery, during which she spent three years battling a resistant infection that threatened her life. Mashood Lawal from Nigeria had a stent inserted into his urinary tract to treat a kidney stone, but developed a resistant infection that nearly killed him; he only recovered when doctors finally found the right antibiotic. In Lebanon, Nour Shamas's mother had surgery on her spinal column and developed a resistant infection of her kidneys, commencing a series of hospitalisations and treatment failures. Gabriella Balasa from the US has cystic fibrosis, meaning she contracts recurrent lung infections which are becoming increasingly difficult to treat because of antibiotic resistance. She lives in constant fear that the next one will be untreatable.

The paradox of antibiotics is that their very success is their downfall: it is as if evolution has a dark sense of humour. The problem of antibiotic resistance has been inherent to antibiotics from the very moment of their discovery, but it has been repeatedly ignored and evaded. To understand the crisis, we must examine where antibiotics came from and how we burned through them.

A history of antibiotics might sound like a rather dry exercise. For most of us, they don't hold great romance or intrigue. When we encounter them they are simply another dull part of modern medicine: white powders, anonymous pills or clear fluids in syringes that we submit to in the hope of getting better. Their names are arbitrary and ugly, often as unpronounceable as they are meaningless. (Drug companies prefer words rich in Xs, Ys and Zs to convey a futuristic buzz, not for any scientific reason.) I have avoided this jargon wherever possible. Behind the pharmaceutical façade lies a sprawling epic spanning all the way from the origins of life on this planet to the cutting edge of modern science.

The story has a diverse cast of characters; not all are human, although the struggles of scientists and doctors to turn their discoveries into medicines have been just as dramatic as the bacterial warfare under the microscope. I begin with the development of germ theory in the nineteenth century, before which humanity was ignorant of the true cause of infectious disease. Each of the following ten chapters is framed around a different antibiotic, moving chronologically through the last hundred years of discovery. As we'll see, each antibiotic has a story of its own: antibiotics have emerged from barrels of coal tar in Nazi Germany, congealed out of mud from the jungles of Borneo and even been predicted by computational algorithms. They range from drugs that have saved millions of lives to new molecules that, at the time of writing, have yet to be tested on humans. By providing the story of such different antibiotics together, my aim has been to produce a group biography – ten threads entwined into a single braid.

This is not an impartial account. I argue that the crisis of antibiotics is, at its heart, a crisis of exploitation. Antibiotics have been developed and used almost everywhere, but though I have searched out stories from around the world I cannot claim this is a truly global history. That is because my aim is to show not merely how antibiotics profoundly changed the relationship between humans and bacteria, but how they helped to forge the modern pharmaceutical industry. To do so, I focus on that industry's centres of power in the twentieth century: Europe and then, later, the United States. I want to explain how an industry that unleashed life-saving antibiotics on the world – and claimed the lion's share of the credit – eventually came to abandon them. History helps not only to explain the reasons for this shift, but also to reveal other ways forward.

In the tale of the sorcerer's apprentice, a young magician takes advantage of his teacher's absence to cast a spell that he does not understand. Commanding a broomstick to fetch water to fill a bath, he rejoices as the spell works perfectly. Only as he watches the water level rise does he realise, with horror, that he doesn't know how to make the broomstick stop. Soon, the whole room begins to flood. Goethe's tale, memorably adapted in Disney's *Fantasia*, is often taken as a parable for the hubris of humanity. Perhaps we are forever doomed to be the arrogant sorcerer's apprentice, unaware of our inability to control the powers that we summon up until it is too late.

It's a beguiling fable. But although we describe antibiotics as magical or miraculous, they operate within the powers of nature – not beyond them. I hope to convince you that bacteria aren't evil spirits beyond our ken, but living beings that evolve and change just as we do. Through studying the history of antibiotics, we can deepen our appreciation for these varied and wonderful beings we share the world with. We can feel our connection with a deeper past. And we can, I believe, take inspiration for the future.

Our ability to comprehend antibiotic resistance at all is a mark of the phenomenal progress that we have made. Throughout human history, most people didn't know that bacteria even existed. At a moment where we can see the problem clearly, it is hardly the time to look away.

Demons under the microscope
Germg theory

The visible world is a late-arriving, overgrown portion of the microcosm.

Lynn Margulis and Dorion Sagan

All societies have recognised that there are two types of disease. Some afflictions are solitary affairs: ''Tis my sole plague to be alone,' wrote Robert Burton in 1621 about his melancholy. Others are transferable, capable of spreading through the population like a stain soaking into cloth. These are infections, named from the Latin *inficere* – a verb that can mean to dye or contaminate. The great Roman doctor Galen wrote that such diseases could cause the breath of sufferers to become corrupted; the vapours they exhaled could then be inhaled by others, engendering the same sickness. The most spectacular realisations of these infectious diseases occurred as epidemics. The Black Death, which began in 1346, may have killed over half of Europe's population in just a few years. To many it would have seemed like the end of the world.

Theories abounded as to the cause of the calamity. Where some believed the plague was due to an unfavourable alignment of the planets, others decided that God was exacting His vengeance on a sinful world. There were also more sinister theories: some said the Jews, a minority long persecuted by the Christian majority, had poisoned the wells. Across Europe, entire Jewish communities were massacred in savage genocidal attacks. It was an age of terror and violence. Around thirty years after the Black Death had first arrived in France, the Duke of Anjou commissioned a vast set of tapestries depicting the story of the Book of Revelation. In vivid threads of red, blue and gold, the Apocalypse Tapestry wove together images of death and destruction that would have been all too familiar to those who had lived through the period. Yet the real cause of this disastrous epidemic was imperceptible, many times smaller than the finest of the tapestry's threads: an invisible bacterium called *Yersinia pestis*.

Humanity would not catch its first glimpse of bacteria until three centuries later. In the mid-1670s, the Dutch draper Antoni van Leeuwenhoek in Delft probably became the first person ever to behold a bacterial cell. Using tiny beads of glass that he held up to the sunlight pouring into his room, Leeuwenhoek was able to penetrate to a layer of reality that was hidden beneath the everyday. He looked at so many things through his glass-bead microscopes, described in hundreds of letters to the Royal Society in London, that it would almost be easier to list what he *didn't* look at. A brief list gives some idea of his vivid and madcap itinerary of the seventeenth-century Dutch world: from wine to frogspawn, tulips to snails. But his most spectacular discovery was that the world was swarming with previously invisible life.

Leeuwenhoek's microscopes conjured living things out of everyday substances, including the plaque he scraped from his own teeth. The beasts he saw there were tiny, but they moved 'very extravagantly'. Different types swam in characteristic patterns: straight lines, loops, corkscrews. (His observations were so accurate that it is possible today to be confident which bacteria he was describing.) These 'little animals' were indisputably alive. Using the mathematics he had learned during previous training as a surveyor, Leeuwenhoek made a simple calculation that there had to be more of these little animals in his own mouth than 'the number of Men in a kingdom'. It may seem strange to a modern view where bacteria are synonymous with disease, but in all Leeuwenhoek's voluminous writings there is no indication that he ever construed the little animals as a threat. If they were apparently everywhere, then how could they possibly cause disease?

For years after Leeuwenhoek, bacteria were seen as benign curios – if they were seen at all. Even in the eighteenth century, the Swedish scientist Carl Linnaeus almost completely ignored bacteria in his system of taxonomy for dividing up the animal and plant kingdoms. Linnaeus's system was meant to bring about order from chaos, but the few bacteria he was grudgingly aware of threatened his scheme. To accommodate their peculiarity meant creating a genus (a taxonomic division above the level of species) literally called *Chaos*. The nineteenth-century zoologist Ernst Haeckel followed Linnaeus in keeping bacteria to a dark taxonomic corner: they were part of the 'formless worms'. Bacteria fell in between the cracks of the two great taxonomic systems of zoology and botany. To some they were closer to animals; to others they were more like plants. Either way, it hardly seemed to matter much.

Into the nineteenth century, most naturalists who looked at bacteria were similarly uninterested. Through the best microscopes, bacterial cells were still only small blobs, as simple as life could get. Even as new ideas about evolution surged through biology like alcohol, illicit and intoxicating, bacteria presented something of a problem. Evolution seemed to mean an increasing complexity of life. But in that case, why were such simple forms of life still around? The French biologist Jean-Baptiste Lamarck thought that evolution was driven by an inevitable tendency towards perfection. Living organisms were climbing a ladder, forever ascending. Bacteria were an ugly blot on this noble idea. Lamarck felt compelled to assert that they must continually arise from nothing by spontaneous generation, spilling out of dirt and putrefaction. How else could one explain the way they clung stubbornly to life's very lowest rung?

Some were less certain. Charles Darwin studied microorganisms like amoebae, and concluded that 'every naturalist who has dissected some of the beings now ranked as very low in the scale, must have been struck with their really wondrous and beautiful organisation'. His point was that initial appearances could be deceptive unless you had the means to look closely. At the time, no microscope could penetrate within the smaller bacterial cell, but experience suggested there was no limit to the microscopic intricacies that evolution had produced. As it would turn out, Lamarck's view that bacteria could not be evolving because they had not turned into animals was a misunderstanding. Bacteria were still bacteria not because they had only recently emerged from nothing, but because they were already exquisitely evolved.

The deep past is dark and shadowy, but it seems likely that for at least 1.5 billion years after life on Earth emerged, the world belonged to bacteria. There was no organism beyond the cell itself: no plants, no animals, no bodies of any sort. Yet life was far from boring. The world that once stood barren and empty was soon crowded, as cells found their way into every crevice they could.

As soon as there was life, there was evolution. All life is based on genetic material stored in molecules of nucleic acids: DNA and its more flexible cousin RNA. In DNA, the script is written in four possible chemicals – by convention denoted A, T, C and G. The pattern of these 'letters' encodes an organism's genome, protected from degradation and decay inside the secure wrapping of the DNA double helix. But every time a cell divides, the double helix must be copied: it is separated into its two strands, then each strand gains a new partner. The two strands of the helix thus become four, one double helix becomes two, and the cell divides.

Every genome in existence has sailed through millions of generations to arrive where it is now, a ship of Theseus with each part copied and replaced countless times. There is no more convincing evidence of the common heritage of all life than the conservation of the basic molecules that carry out these processes. Right now, enzymes are at work within your cells that share stretches of sequence identical to a similar enzyme in *Escherichia coli*, a bacterium that lives within the human gut. Such molecular motifs are rhyming couplets across the great poem of life. Whereas at other points evolution may extemporise, conjure great loops of digression or hit on a new turn of phrase that exceeds the original, there are some passages that cannot be changed.

Just as an organism's genome passes on its accumulated wisdom to its descendants, so human societies have sought to safeguard their knowledge against the ravages of time. The outcome of just one of these endeavours shows the impossibility of that task. In the thirteenth century, the Goryeo dynasty of Korea commenced a grand undertaking. The three canonical collections of the Buddhist canon, known as the Tripiṭaka, would be engraved onto white birch, wood so dense that it sinks, to be preserved for posterity. Logs of birch were soaked in seawater for three years and then cut into thousands of blocks, each one the weight of a newborn baby, which were then boiled in salt and left to dry outside for a further three years. Only then were their surfaces ready for carving. For over a decade, craftsmen carved more than 52 million characters into the blocks, reciting a prayer after each character was finished. Today, the Tripiṭaka Koreana is stored in the Temple of Haeinsa, stacked up on shelves beneath elegantly curving roofs, nestled in the green foothills of Mount Gayasan. The low wooden buildings, protected by full-time security guards and a fire truck stationed within the grounds of the temple, house 80,000 individual blocks, each one a pure repository of perfect knowledge.

At least, that was the intention. Modern scholars have found that the Tripiṭaka Koreana, for all its breathtaking beauty, contains hundreds of mistakes that presumably arose from the everyday chaos of life – a smudge of ink, a misreading, or the slip of a craftsman's knife.

A bacterial cell has much the same problem. Its genome must be copied every time it divides, which in its most vibrant growth phase may be as often as every twenty minutes. *Escherichia coli* has over 4 million characters of DNA in its genome. A long-term experiment established that, on average, in 1,000 generations of copying and recopying between cells, its genome had accumulated less than a single difference. Every twenty minutes, a tiny bacterial cell in your gut succeeds in doing what the Goryeo dynasty could not. But still, the transmission of information is not perfect. A mutation is an event where the letters of DNA change randomly. In terms of fidelity to the past, any mutation is an error, yet the unexpected beauty is that sometimes, by chance, it presents an improvement. It is this fundamental tension between order and disorder that is essential to all life.

The first life consisted of single cells. Yet it was impossible for these cells to be sealed units. Life is a process: without an ebb and flow of chemicals across a cell's outer boundary, it would die. As chemicals flowed, the first bacteria began to evolve new networks for biochemical reactions, reaching new areas of molecular possibility. As these new molecules proliferated, the unwanted output of one cell could be the input of another. Simple pass-the-parcel exchanges between cells flourished into byzantine arrangements of interdependence and competition. As the microbes multiplied and diversified, they constructed a complex web of alliances and rivalries, like a sprawling medieval war. As bacteria banded together or floated free, they differentiated into separate species, each adapted to a different set of molecules and environments. The concept of a biological species as developed by Darwin and others had been one of reproductive isolation: each species was a delicate and separate branch on the great tree of life. But that concept does not apply to bacteria. The implications are mind-bending. 'The entire bacterial world', writes the doctor and scientist Stuart Levy, 'can be thought of as one huge multicellular organism.' From life's earliest moments, genes were being carried by viruses and other rogue genetic elements between cells, shuttling back and forth as if on a molecular loom.

Within this chaos, bacterial species could still form, but they had to carve out a space for themselves. Given this, the evolution of chemical weapons was inevitable. Bacteria made molecules that poisoned their enemies and spared their allies. Such molecules were the first antibiotics. Soon, evolution produced in turn molecules that could defend against antibiotics. These developments were less sequential than this description implies; they must have unfolded in chaotic parallel. Resistant bacteria evolved and prospered wherever antibiotics did. The two were locked in an ecological balance: a rise in antibiotic production by one bacterium would drive a rise in resistance in another. Antibiotics and antibiotic resistance swirled around each other like yin and yang.

However, this single-celled world did not last. Around a billion years ago, before most of the stars visible in our night sky had started shining, something strange happened. The cell which had served bacteria so well underwent a sudden change, creating a new form of life: the eukaryotes (*yoo-KA-ree-ohts*). Most biologists believe that the origin of eukaryotes was spectacularly improbable – perhaps a once-in-Earth's-lifetime event, with a fusion of two very different types of single-celled organism: archaea and bacteria. Explaining exactly how that happened remains a mystery – the biologist Nick Lane has called it 'a black hole at the heart of biology' – but somehow, it did. The new eukaryotes were fundamentally different from their ancestors. Their name comes from the Greek for 'well-formed nucleus', denoting their altered cellular structure. The genome no longer floated free in the cell but was bundled in a dense knot – a nucleus – like the yolk within an egg. The evolution of the eukaryotes marked an important milestone in evolution. Their cells were a complex fusion, and complexity began to develop at a level beyond the cell too. For reasons that are still debated by scientists, eukaryotic cells did something no bacterium had ever done and flourished into multicellular life, creating the first bodies. An organism was now no longer limited to a single cell but became an extended piece of real estate. A single being could contain multitudes.

The existence of multicellular organisms did not spell the end for bacteria; it presented an opportunity. The first bodies were rich concentrations of nutrients, sequestered into one space and guarded against the hungry microbial world with barriers of shell, skin or mucus. When a multicellular organism died, that imbalance was repaid. As the body decayed and broke down, microbes would flood in, reconfiguring the molecules of the body into new bacteria. Out of the raw materials death provides, bacteria could decompose a new symphony of life.

But bacteria didn't need to wait for death. Like a cell, a multicellular body itself needed to be permeable, making it impossible to keep them out entirely. One of the first challenges for multicellular organisms was keeping that hungry world at bay. Organisms allowed some of the bacteria in, forming ecological systems within early animals: the first microbiomes. Nascent immune systems evolved to control and even select these resident bacteria. Such bacteria could provide a multicellular organism with benefits, including that simply by occupying space they prevented more harmful bacteria from gaining a foothold. As eukaryotes continued to evolve, the bacterial species that gave up their freedom to live inside these increasingly complex bodies were a minority. But they were a concentrated minority. The mammalian gut created perhaps the richest assemblage of bacterial diversity in life's history, where hundreds of different species coexisted in close proximity.

Bodies came with other challenges. One organism could now contain many different types of cell, the same genome growing into an unruly menagerie that somehow pulled together in the same direction. Your body contains about 200 different types of human cell. From the tiny saucers of your red blood cells to the metre-long neurons that stretch along your spine, you are a teetering assemblage of difference bound together by natural selection. The same genome gave rise to all of these cell types, and each is regulated and balanced against the others in a tight social contract. Your body collectively maintains homeostasis, balancing conditions and resources to ensure continued survival of the total assemblage. A healthy cell obeys its orders and divides in a regulated way, even killing itself on demand if it receives the signal. But that social contract can sometimes break down. In cancer, a mutant cell just keeps doubling as if it has gone mad and forgotten its responsibilities: one, two, four, eight, sixteen . . . In twenty doublings there will be more than a million cells. Left unchecked, this rogue quest for infinite growth wreaks havoc on a finite body. Siddhartha Mukherjee has called cancer 'a pathological mirror' of our own existence. Its cellular atavism of uncontrolled growth is a return to an ancient state of life, the purest evolutionary expression of selfishness.

Bacteria cannot get cancer: they already have it. As single-celled organisms, they have not evolved to coordinate a complexity of cell types. While the good news is that no bacterium wishes you harm, the bad news is it may kill you anyway if its mindless passion for growth overwhelms your body's normal functions. To take one example, consider the rare bacterial condition known as infective endocarditis. As Leeuwenhoek observed in the seventeenth century, the human mouth contains many different bacteria growing on the gums and teeth. Normally, these bacteria pose no serious threat, as long as they are regularly removed by the brushing of teeth. But if these bacteria are given access to the bloodstream – say, through the removal of an abscessed tooth by a dentist – they can move elsewhere within the body. In rare cases, certain species can be pumped around the bloodstream and back to the heart where, if they are lucky enough to cling to the surface of a heart valve, they can grow and grow, forming a clump of infection analogous to a tumour. The heart is compromised: perforation or rupture is possible. Chunks of the infection may detach and float further through the bloodstream, causing chaos wherever they lodge in the body. As in cancer, in a body of many cell types the unchecked growth of just one can wreak havoc. Worse, some bacterial species manufacture their own toxic poisons to assist with their conquest, and have evolved to spread not just within but between human bodies. Bacterial infection is a form of single-celled revenge against the complexity of multicellular life.

The fact that bacteria can cause disease is so ingrained in us now that it is difficult to appreciate how ludicrous the idea of germs was when it first emerged. In the nineteenth century, cholera began to spread throughout the world, causing devastating outbreaks. Sufferers found themselves racked by severe diarrhoea and vomiting – the word 'cholera' may have arisen from a Greek word meaning 'gutter'. Doctors were agreed that cholera was infectious. As the British doctor John Snow wrote, 'It travels along the great tracks of human intercourse, never going faster than people travel'. Where they disagreed was the cause. Many continued to follow Galen and blamed bad air, but Snow thought otherwise. In 1854, he studied a cholera outbreak in the Soho neighbourhood of London, the worst that he had ever seen. Drawing up a map of cases and enquiring locally, Snow homed in on a single culprit: the water pump on Broad Street.

But when he first examined the water for contamination, he could see almost nothing. 'I found so little impurity in it of an organic nature,' he wrote, 'that I hesitated to come to a conclusion'. He was certain that the water was to blame, but he couldn't explain why.

Leeuwenhoek had seen little animals in water almost two centuries before. But how could little animals be responsible for disease if they were present almost everywhere – and in perplexing diversity? As the French philosopher of science Bruno Latour put it, falling sick from some invisible corruption was 'all the more disturbing' because it did not always happen. The enemy's lair had been identified, but the enemy's true identity remained as murky as the clouded water.

The only way that most people could see bacteria was in the spoiling of food and drink, and such a general form of corruption could not be attributed to a single species. It was only when scientists began to catalogue and enumerate bacterial diversity that they realised not all bacteria were alike. At first, mimicking the decay of food, bacteriologists tried to grow individual species on solids like meat, potatoes, eggs or bread. But the growth was erratic and too many different types grew together. The trick was to appreciate that bacteria prefer their food easily digestible – a soup of raw inputs – and to use liquid broth. Scientists became providers of bacterial *haute cuisine*: 'One obtains a culture medium by leaving for twenty-four hours, in contact with twice its weight in water, finely chopped lean veal.' The aim was to grow up pure populations of single species in order to study them in isolation.

To isolate a single bacterial species was an extraordinary achievement – a purity never seen in nature. Scientists began with broth sterilised by heat, as devoid of life as it was full of nutrients. They then impregnated the broth with the smallest amount of matter imaginable: a single bacterial cell. The French scientist Louis Pasteur praised the virtues of using a needle-thin glass rod as a means of transfer, so sharp its point was barely visible. Once in the broth, the cell would divide again and again, until the flask that had once been lifeless was teeming with unseen millions.

As further epidemics of cholera had raged through cities in the 1880s, it was in Kolkata in India that the German bacteriologist Robert Koch identified a comma-shaped bacterium as the real culprit: *Vibrio cholerae*. Like many bacteria, it could float unseen in water, but it alone was responsible for the devastating illness. If swallowed, it could multiply in the sufferer's gut before exiting in their diarrhoea, contaminating more water and infecting fresh hosts. One doctor who didn't believe Koch's claim was Max Pettenkofer. Pettenkofer had founded Munich's Institute of Hygiene and redesigned the city's sewers to address previous epidemics, but he disagreed that bacteria were the only important factor. To prove it he wrote to Koch and requested a pure liquid culture of *Vibrio cholerae*; when it arrived he drank it down. Fortunately, Pettenkofer survived. It's thought that Koch was aware of his plans and deliberately sent him a weak concoction, preferring to lose the argument than be accused of murder.

There is a strange intimacy between enemies. As work from Koch, Pasteur and other microbiologists became more widely accepted, the 'germ theory' became the dominant explanation for infectious disease. Bacteria had been transformed from Lamarck's hapless blobs to demons under the microscope. The task of bacteriology was to examine them in the laboratory and work out how to defeat them. But though isolating single species was a spectacular feat, to the naked eye bacteria could still manifest only as cloudy liquids. Though contaminated broth was fertile with bacteria, in a visual sense it was barren – a medium of monotony. It was Koch who gave a vivid demonstration that turned the world of bacterial culture three-dimensional. Once again, it involved borrowing from cookery. He mixed broth with animal gelatine, the same substance used to make jellied desserts. Koch's jellies were translucent reds and browns brewed from blood or beef. The resulting patch of jelly contained the nutrients the bacteria needed to grow on a flat surface. What Koch called a 'moist chamber' would come to be known, after refinement by his assistant Julius Petri, as the Petri dish.

When Koch demonstrated the technique in London at an

international conference in 1881, people were astonished. Here one could *see* bacteria, plain as day. In one demonstration, an empty dish was left with its lid off outside in the London air for four hours. Koch then brought it back inside to show to spectators. Forty or fifty different spots had formed of varying colours, each corresponding to the slowly spreading circle of a single cell that had landed there by chance. As one spectator put it, it proved that the air itself contained species with different characteristics – what had been invisible now flourished 'visible to the naked eye'.

The gelatine on the plate was later replaced by another substance using a suggestion from Fanny Hesse. Hesse, known as Lina by her family, was married to Walther, a microbiologist who worked with Koch. She herself worked with Walther as a microbiologist and suggested an alternative to gelatine. Hesse had grown up in New York, where from Dutch neighbours who had emigrated from Java she had learned of agar, a substance made from seaweed extract. Agar, a clear jelly with almost no taste, could be used as an alternative to gelatine in cookery to make set puddings. It proved to be even better than gelatine for bacterial cultures and remains at the heart of almost all bacterial research.

Koch and his colleagues had found a middle way between the figurative but intangible world of the microscopic cell, and the tangible but abstract world of broth. Just a few years earlier, Claude Monet and his fellow Impressionists had shocked art critics by giving them only an 'impression' of a scene, rather than attempting to reproduce it. Similarly, bacteria could now be understood through the 'impressions' they left behind in the lab. A single species could be painted onto a blank canvas for closer study, its opacity, pigmentation and texture on the dish becoming part of its identity. A language of streaks, circles and discs became part of the visual lexicon of science. Once bacteria became visible, they could not be unseen.

Cholera was not a fluke. As bacteriology grew as a discipline, one by one some of the world's deadliest infectious diseases were unmasked as having a bacterial cause. Unfortunately, this knowledge didn't seem to translate into effective medicine. There were public health improvements, but these were mostly thanks to better sanitation and living conditions of the sort advocated by Koch's intellectual adversary Pettenkofer, as well as better nutrition. In Britain in 1800, 329 in 1,000 children died before the age of five. By 1900 the figure was down to 228, and by 1930 it would more than halve again. As Latour put it, bacteriology was 'far from being the source and cause of hygiene' – it was 'merely a ripple on the surface'. Prevention, not cure, was the way to save lives.

The few available treatments for bacterial disease used other bacteria. Once bacteriologists could isolate single species, they had been able to combine them, trying to build back from the purity of a single species to the messiness of the real world. Applying Darwinian ideas about evolution to microbes, bacteriologists had become intrigued by competition between different species of bacteria. Could these antagonisms be exploited for human benefit? In the late nineteenth century, scientists had discovered that the bacterium known as *Pseudomonas aeruginosa* could kill pathogenic bacteria. Unfortunately, it was itself a deadly pathogen, associated with a characteristic blue pus that oozed from infections. This dirty contaminated broth was hardly a reliable medicine. Eventually, a single molecule called pyocyanase was isolated from the blue broth, but its use was limited to eye-drops, spray or mouthwash because swallowing it was toxic.

Another approach used animals, not bacteria, to make medicine. Scientists had found that bacteria often released toxins that killed human cells. In 1890, the Prussian scientist Emil Behring and the Japanese bacteriologist Kitasato Shibasaburō discovered that injecting horses, guinea-pigs and goats with toxic bacteria allowed 'antitoxin' to be obtained from their blood serum. The antitoxin was a chemical that bound to the bacterial toxin and so, if injected into other animals, reduced the severity of the disease, known as diphtheria. Here was medicine manufactured by an animal's own immune system. Applying the same principle to cholera also yielded a similar treatment. Two important bacterial diseases now had treatments, even if they only diminished the virulence of the disease by mopping up toxin rather than directly attacking the bacteria. To manufacture diphtheria antitoxin, the most efficient method was injecting horses with bacteria, then extracting their blood and giving it to children. Such was the need that public health departments in America acquired their own stables. At one stables in New York, a seven-year-old bay gelding was bled nine times in a year, producing twenty-two bottles of serum.

Antitoxin serum and blue broth were attempts to bottle living elixirs. They worked, but they were messy magic. Would humanity need a different stable for every disease?

At the start of the twentieth century, some felt that searching for medicines in biology was misguided. Instead, they perceived a gleam of hope in a different scientific realm. Rather than trusting in living organisms, these scientists saw in the lifeless mysteries of chemistry a new and toxic possibility. In their test tubes they aimed to brew poisons of unprecedented power.

The red thread

Prontosil

Without the thought, though the material parts already exist,
the form does not and cannot.

Dorothy L. Sayers

The Industrial Revolution arguably began around 360 million years ago, when a tree fell in a forest. Nobody was there to hear it, but we know the sound it would have made: not a crash, but a squelch. This was a time when the world was covered with swampy forests, filled with strange and spiky trees. Horse-tails jostled with lycophytes as dragonflies droned lazily through the muggy air.

The tree that fell was a fern. Rather than branches, a frothy bouquet of green fronds emerged from the top of its stubby trunk covered in matted brown hair. The trunk was made of wood; the wood was made of cells; the cell walls were made of tough molecules called lignins, consisting of oxygen, hydrogen and carbon linked together in scaffolds. To make them, the tree had used its green fronds to bottle sunlight into molecular energy through photosynthesis. The lignins had allowed the tree fern to stand tall. For trees that fall in a forest today, bacteria and fungi begin to rot the wood and break it down. But that wouldn't happen to this tree.

As it lay on the marshy ground, it was soon submerged. There was little oxygen in the soupy water, suffocating most processes of decomposition before they could take effect. The tree was sequestered away from the cycle of organic matter, while above it the forest kept growing. More trees lived and died; more trees fell, building up in layers. The earth was full of ancestors, and the weight of the forest's present bore down on its past.

The past began to change. Once the tree was deep underground, buried beneath its descendants, its world became one of heat and pressure, and it began to be transmogrified. Molecules were shifted and squeezed, their atoms reconfigured into new arrangements. The transition was imperceptibly slow, but there was plenty of time. The continents which had once coalesced like a blob of mercury into the supercontinent Pangea eventually broke apart again and floated around the surface of the globe, taking their layers of ancient forests with them.

Just under 300 million years later, hominids evolved. In time, a species called *Homo sapiens* would dig into the earth, finding strange layers of flammable black rock. The journey from plant matter to these layers had been arduous. The first stage in the process was peat: a black soil that when dried out burns with a delicious smoky smell. The sea had then come creeping over the land. Marine sediment fell gently over the peat, covering it in a soft blanket and forming a stratum of matter that travelled deeper and deeper underground as the layers above accumulated. Given more pressure, more heat, more time, the peat had hardened into lignite, a crumbly brown rock that burns with a choking flame. Compressed still further, it formed jet, a black, lightweight stone. In the first century AD, the naturalist Pliny the Elder noted the resemblance of jet to wood, writing in the thirty-sixth book of his sprawling natural history that it could be burnt to ward off snakes. Placed on the head of a hatchet and heated, it could reportedly prophesy the future. But jet was not the final stage. As the tree fern was compressed into a smaller and smaller space, it metamorphosed further. The sprawling molecular jigsaws of the ancient lignins grew harder and blacker as oxygen and hydrogen were squeezed out and the carbon density increased. The tree fern had turned into coal.

Almost all the world's coal reserves formed between 360 and 300 million years ago in a period known as the Carboniferous, literally meaning 'coal bearing'. Prehistoric coal could serve as the fuel of the

Industrial Revolution because it was effectively concentrated wood, with an energy density 50 per cent greater. Coal burned quickly and purely, whether in a hearth or in a steam engine, as energy that had accumulated for millions of years was released in one final blaze of glory. Burning ancient hydrocarbons – and in the process, producing carbon dioxide and heating the world's atmosphere – became the world's main source of energy.

From the heady brew of hydrocarbons, new chemicals emerged. By studying them, chemists began to understand the chemistry of living things. After millions of years underground, the Carboniferous forests were about to embark on a strange afterlife. From ancient forests, chemists would conjure radical new medicines that killed bacteria. Though these molecules were not made by microbes, they were miracle drugs: the first antibiotics.

The forest had always been a source of medicine as well as fuel. From herbal salves to hallucinogenic mushrooms, it was clear that certain plants could have considerable effects on the human body beyond being food. Yet plants were fickle pharmacists: where they grew was often maddeningly inconvenient. And the forest itself could be deadly. In the nineteenth century, Europeans who travelled into tropical regions frequently contracted malaria, a debilitating disease that caused intense chills and fevers, and in some cases death. Malaria had once been endemic in marshy parts of Europe, but after the draining of swamps, most Europeans had rarely encountered it and so their immunity was low. The cause of malaria was only identified in 1897 by the British doctor Ronald Ross as the *Plasmodium* parasite – a microbe, but one more closely related to humans than bacteria – that was transmitted by mosquito bites. But looking back over the nineteenth century, Ross accurately summarised the views of previous colonial scientists when he described malaria: in his eyes it was 'a gigantic ally of Barbarism' that had thwarted the expansion of European 'civilization'.

While the precise cause of malaria was unknown before Ross, there had long been a tentative treatment. Jesuit missionaries who had travelled to Peru in the seventeenth century had discovered from the indigenous people they met there that a particular tree could serve as a medicine. The tree – it was more of a large shrub – grew only at high altitudes, often in inaccessible places. It had narrow green oval leaves and bunched fingers of flowers that could be white, pink or red. But its value lay in its bark. In 1630, the bark was supposedly used to treat the Countess of Chinchón in Lima, and the common name for the tree became cinchona.

Cinchona was a useful if unreliable passport for Europeans – consuming its powdered bark sometimes treated malaria, but success rates varied. More inconveniently, the bark could still only be obtained from a small region of Peru and Bolivia. Attempts to bring back cinchona to Europe proved largely unsuccessful. The plants didn't travel well, or if they made it back they did not thrive.

But in 1820, two French chemists studying cinchona isolated a new chemical. They named it after the word for the tree in the Quechua language: quina became quinine. Quinine was the active ingredient in cinchona, and could treat malaria on its own. No longer did the cinchona bark itself need to be consumed by the patient. It was the first time an isolated chemical from nature had been used against an infectious disease.

On the other side of the Atlantic to the cinchona trees, European nations wanted powdered quinine to facilitate their land grabs in Africa. There was already a far greater demand for quinine than could be satisfied by the existing supply. What if the drug could be synthesised using chemistry? Though quinine's exact chemical structure remained a mystery, it was known that it consisted of just four elements: carbon, hydrogen, nitrogen and oxygen. To some chemists, quinine was beguilingly close to the hydrocarbons produced from burning coal. A German chemist speculated that a 'happy experiment' could make it in the laboratory. In 1856, the eighteen-year-old William Perkin tried and failed – but in the process found something remarkable.

Perkin set up a laboratory in a small room at the top of his father's house in the East End, with a view of the sprawling dockyards where, every day, ships arrived from every corner of the British Empire. His experiments to produce quinine failed, producing only a black sludge. But rather than discard the sludge, he purified and dried it to get another powder. This powder wasn't black but a rich purple: mauve. Testing it by staining a silk cloth, he found that the fabric took up and held the mauve colour easily. He called his new compound mauveine.

Purple was a difficult colour to come by. Like quinine, it came from natural products, and little in its production had changed in nearly 2,000 years. In the ancient world, Pliny had written about purple dyes. One source was a species of sea snail, which when ground up was used to make Tyrian purple, the colour of emperors. Another was the indigo plant from India. Pliny wrote that this 'slime that adheres to the scum upon reeds' could yield, when diluted, a beautiful mixture of purple and sky-blue. Even in the nineteenth century, such natural dyes were still economically important: the British Raj forced Indian farmers to grow indigo rather than food crops, then bought it at artificially low prices. Now mauveine presented an alternative.

The revelation of Perkin's mauve was followed by other synthetic dyes from coal-tar products; these molecules began to force natural dyes out of the market. Fashion exploded from the muted tones of colours squeezed from living organisms into outrageous new hues. The artist and designer William Morris despised these dyes, calling them 'one of the most wonderful and most useless of the inventions of modern Chemistry . . . a series of hideous colours, crude, livid—and cheap,—which every person of taste loathes'.

Others disagreed. In 1862, at the London International Exhibition, there was a dazzling display of a rainbow of dyed fabrics. In the midst of all this beauty was its ugly source: a sample of coal tar, black and repulsive. Concentrated chemistry produced a new language of colour: Acridine orange and Alcian blue, Bismarck brown and rose bengal, Congo red and crystal violet. These names were false pedigrees; the chemicals had emerged from the laboratory and had no real connection to the evocative names bestowed on them.

By 1900, a single German firm was synthesising the same quantity of indigo dye from coal tar as could be obtained from a quarter of a million acres of Indian farmland. Though it had been synthesised by chemists through complex reactions, the raw material at the molecular level came from the vegetation of long-vanished Carboniferous forests. And the vibrant expertise that developed in the dye industry was soon soaking into other disciplines too, staining them indelibly.

Microscopes had revealed that human bodies and bacteria alike were made up of cells, and it seemed obvious to try to understand disease by understanding how these fundamental units of life interacted. But both human and bacterial cells were transparent and largely colourless. Trying to work out their interactions through observation was a little like trying to become a car mechanic by gazing into a three-dimensional engine made entirely of glass. But in the late nineteenth century, scientists found that the new chemical dyes could selectively stain different cell types. Now, it was as if each cog in the engine could be dyed a different colour.

One of these scientists was Paul Ehrlich, a German workaholic who seemed to subsist on cigars and mineral water. He approached science as detective work, calling it 'the method of Sherlock Holmes'. A signed portrait of Arthur Conan Doyle hung on the wall of his house. Ehrlich was an intensely visual thinker who would draw on any available surface. He adored bright colours and could become ecstatic over a bunch of flowers. In particular, he loved dyes. In the 1880s, he showed how they could be used to distinguish bacterial and human cells. In sputum from a patient with tuberculosis, the bacteria causing the disease would show up brightly under the microscope. One could now easily diagnose tuberculosis as distinct from other lung infections. Synthetic dyes rapidly lit up the microscopic world.

In 1884, the Danish bacteriologist Hans Gram was working in the morgue of Berlin's city hospital. Like Ehrlich he was trying to diagnose disease under the microscope. His samples were of lung tissue, obtained from the corpses of people who had died from pneumonia. Pneumonia leads to inflammation and damage to the lungs, and Gram found that large volumes of blood had frequently leaked out into his samples, making the bacteria impossible to see. Inspired by Ehrlich's techniques, he used a dye called gentian violet in combination with a solution of other chemicals to stain the whole sample purple. He then attempted to wash out the colour with alcohol, in effect rinsing the purple away. However, the dye remained stuck onto the bacteria, which then showed up as an intense colour against the faint-yellow canvas of the human tissue.

Gram noted, almost in passing, that some other types of bacteria did not appear to take up the dye. This offhand observation came to offer a way of classifying bacteria into two broad types: those that took up the purple dye and those that didn't. These would become known as Gram-positive and Gram-negative. (The simple division would eventually become a crucial one for antibiotic development: the same factors that made Gram-negative bacteria difficult to stain would turn out to make them much harder to get antibiotics into.) Dyes were a new way of seeing, revealing unseen order within the bacterial world.

In 1898, Ehrlich became the first director of the new Royal Institute for Experimental Therapy in Frankfurt. A bequest from Franziska Speyer, the widow of a wealthy Frankfurt banker, was used to build a brand-new building next to the institute. The Speyer-Haus would be owned by the city of Frankfurt, but run as a private institution. Ehrlich was made its first director in 1906.

German writers speak of *der rote Faden* – the 'red thread', the central concept that runs through a work or a life. Ehrlich's red thread was that dyes were proof of binding at the molecular level. And if dyes could be selectively bound to bacterial cells but not human cells, what would happen if the dye was also a poison? He felt that the logical basis for treating infections should be to develop a way of delivering poisonous substances to only certain types of cell. After all, this was what the immune system itself did through the production of antitoxins. Though what antitoxins did within the body was dark and unseen, as a chemist, Ehrlich had a characteristic visual metaphor to hand. For molecules to interact, he reasoned, they had to bind to each other, and so there must be a fit between the 'side-chain' of the antitoxin and a corresponding receptor on the toxin: the antitoxin slotted in like a key in a lock. Ehrlich's antitoxin theory won him the Nobel Prize in 1908. But he dreamed further and more extravagantly. Riffing off the idea from German folklore of a magic bullet which always hits its intended target – a *Zauberkugel* – Ehrlich suggested that the immune system as a whole worked in a similar way. He visualised magic bullets made by the immune system rico-cheting around the body, bouncing harmlessly off everything until they finally found their targets. That fantastical image suggested another possibility. Rather than relying on the body's magic bullets, why couldn't chemists manufacture their own?

The concept of using poison against diseases was not new. Toxic compounds like mercury had been popular in medicine since the seventeenth century. One of the diseases they treated was syphilis, a common sexually transmitted disease that could also be passed on in pregnancy. At the start of the 1900s, around one in ten adults in England had the disease. Syphilis was 'the great imitator': the disease could progress through a bewildering perplexity of symptoms, from blotchy rashes to slurred speech if the bacterium infected the brain. The shapeshifter's true form was identified in 1905 as a corkscrew-shaped bacterium; poisons could harm the bacterium, but could kill the patient too. Ehrlich decided to modify the poison to try and produce a magic bullet. His programme started with arsenicals – compounds that contained arsenic – and sought to make them less poisonous. Arsenic is murder to most organisms, bacterial and human alike. Could its deadliness be retained for bacteria while making it sufficiently safe for animals?

Working in Ehrlich's institute, the Japanese scientist Sahachiro Hata tested a range of different arsenicals on rabbits with syphilis, and found that one of these compounds, referred to as 606, could cure the disease. Ehrlich was vindicated. On 19 April 1910, in front of a medical audience, he announced the existence of a cure. But 606 wasn't yet being manufactured by a company: it was only available from the Speyer-Haus. Ehrlich ordered the chemistry department to drop everything else and only make 606. Physicians made pilgrimages to Ehrlich from around the world, hoping to get a sample of the new drug. There were so many that seeing Ehrlich himself could mean waiting for days. The atmosphere was frenzied – some people became almost violent.

Ehrlich was excited. But he was also nervous about the possibility of sending out faulty batches – he personally inspected each one. By September 1910, the Speyer-Haus had dispatched 65,000 free doses, with the expectation that physicians would send back reports directly to Ehrlich. (The concept of a clinical trial did not exist, and the experience of individual doctors was the only way a medicine could be judged.) Ehrlich's oversight was fanatical but idiosyncratic: he kept a list of recipients, doses and dates by writing in pencil on the inside door of a bookcase in his study. Once physicians got the drug, 606 was not simple to administer. It came as a yellow powder that needed to be prepared into a liquid for injection. To start with, Ehrlich recommended mixing it with distilled water and caustic soda, then injecting this solution into the patient's buttocks. It was horribly painful; many patients needed morphine to make it through. Ehrlich frequently revised his instructions as clinical reports came in, sending urgent notices to medical journals. When it became apparent that distilled water obtained from pharmacies was frequently teeming with other bacteria – pharmacists kept the water in big jars, where it sat for days on end – he emphasised the need for fresh water. His fear was that 606 would be blamed for complications of infection caused by unsterile injections. In October, he switched to recommending that injections be done directly into a vein in the arm, which was a significant hurdle for most doctors because at the time it meant a surgical procedure.

In December 1910, large-scale production of 606 began. It was branded as Salvarsan: the arsenic that saves. Indeed, it saved. But it wasn't a magic bullet – it was more like a cluster bomb. The arsenical might have been modified, but it still contained arsenic, toxic to both bacterial and human life. But when compared to the dearth of other effective treatments, Salvarsan was, on balance, worth it – in the views of many doctors, it was 'the most remarkable synthetic chemical compound introduced into medicine'. In this sense, it was a realisation of Ehrlich's dream. But it was a trade-off. Doctors had to inject just enough to kill the bacteria without also killing the patient. Salvarsan could, and did, kill people. It could exhibit a miraculous cure if delivered very early in the progression of syphilis, but many patients would relapse, and it could take up to forty doses administered over more than a year to enact a cure. And for the later stages of syphilis, Salvarsan couldn't help. Nevertheless, its existence was transformative.

The Salvarsan licence started to deliver profits almost immediately, 55 per cent of which went to the Speyer-Haus where Ehrlich worked. As a result, by 1911, the institute's budget had already increased. Ehrlich had the drug manufactured by Hoechst AG, a pharmaceutical company that he had been involved with since its days as a producer of dyes. By 1914, Hoechst AG was shifting 6.6 million marks' worth of the drug in Germany (equivalent to around $50 million in 2024). The amount of money rolling in led to accusations of profiteering. Ehrlich was Jewish, and this became a focal point for a grotesque antisemitic caricature: the Jew who enriched himself at the expense of others' suffering. After the outbreak of WWI, Ehrlich died in 1915 – according to his friends and colleagues, of exhaustion from the attacks against him and Salvarsan.

Physicians around the world were now desperate for Salvarsan, but few could get it. The drug showed how important the intertwined dye and pharmaceutical industries had become. Unlike a natural medicine like quinine, where supply relied on harvesting a tree, Salvarsan could be manufactured synthetically in unlimited quantities – but that expertise was concentrated in Germany with its advanced dye industry. And then there was the question of patents. Who owned Salvarsan?

Up until the late nineteenth century, the concept of intellectual property had been largely irrelevant to medicine. Most medicines didn't work, so ownership of them was a moot point. Those that did, like quinine, tended to be relatively simple and made from natural ingredients. No one would dream of trying to patent a tree, and claiming exclusive rights to a medicine was widely considered unethical. But effective medicines manufactured by chemistry represented something new.

In Germany, the dye companies had been particularly concerned with patent laws, using their power to influence the system. The Patent Act of 1877 had established a system of patents where, after examination for novelty, a patent would offer protection to its inventor for fifteen years. The process was unlike other countries such as Britain, where patents were simply registered and not examined. Under the German Patent Act, patents were awarded to the first inventor to file, and renewal fees were high; both these rules meant that large firms were at an inherent advantage. But the Act had limited scope: it specifically excluded foods, drinks and medicines. Even for chemicals, it was only possible to patent the manufacturing process, not the chemical itself. Dye companies had learned to strike deals, agreeing not to challenge patents held by each other while attacking rival firms who tried to make similar products.

However, that was the situation inside Germany. A patent held in Germany wasn't necessarily respected abroad. In the view of German firms, Switzerland was the 'pirate nation'; France was the 'country of counterfeiters'. So where possible, German companies also filed patents in foreign countries. In the US, the dominance of German firms in chemical patents was overwhelming. In 1912, the US Tariff Board found that 98 per cent of patents related to chemicals were assigned to German firms and 'were never worked in the United States': they had been filed only to protect the firm's interests, not as a prelude to manufacturing. The holder of a patent could allow other companies to use it under licence – but there was no requirement for them to do so.

These concerns came to a head with Salvarsan. The war had

stymied international trade with Germany after August 1914, when the British had enforced a naval blockade of German ports, drastically reducing the supplies of Salvarsan to countries like the US. The patent rights for Salvarsan in America were tightly controlled by one man: Herman A. Metz, an American businessman, who was the authorised agent of Hoechst AG. The son of two German immigrants, Metz was a powerful political figure in his own right, a former congressman and one-time Comptroller of the City of New York. Now, he was the only person who could grant licences to import and manufacture Salvarsan. Metz set the price at around $4.50 a dose, but a black market in Salvarsan boomed because of the blockade: at one point in 1915, the price of a dose rocketed to $35 as demand outstripped supply.

A small group of American businessmen saw a commercial opportunity and funded the construction of a vessel to transport goods across the Atlantic, circumventing the blockade. Not a ship – a submarine. The *Deutschland* set off from Heligoland in northern Germany into the North Sea, diving underwater for ninety miles to evade the British ships, then skirted round the northern coast of Scotland before heading across the Atlantic, arriving in Baltimore in July 1916. It was loaded with the most valuable cargo possible: German dyes and pharmaceuticals. The *Deutschland* made a second trip later in the year. Delivering Salvarsan by submarine was as lucrative as it was audacious, but a single submarine wasn't going to solve the supply problem.

What was needed was increased manufacturing in the US. After the US entered WWI in April 1917, the patent situation increasingly rankled. As the physician Dr Walker complained in an open letter written with a colleague in May, the drug had been patented by Germans working for a German company; the US patents mentioned 'no American name'. Walker estimated that the drug was being sold at around ten times its cost price. His demand was simple: he wanted the US government to rip up intellectual property rights in this specific case. He called on the government to declare the Salvarsan patents null and void.

A proposal to abrogate the Salvarsan rights was put before the Senate Patents Committee on 4 June, and Walker attended to give evidence. When he stepped outside into the corridor, Metz was waiting for him. As Metz approached Walker, he swung his arms – 'either by accident or intention', in the words of one witness – and his outstretched palm brushed against Walker. In response, Walker punched him in the chin, throwing Metz backwards against a wall. An attendant ran in and stood between them to break it up. After the physical skirmish, Walker eventually won the legal battle too. In October, the US government abrogated the patents: American firms could now make Salvarsan.

The Salvarsan patent crisis made it clear how crucial the chemical derivatives from fossil fuels had become. The dye industry was the foundation for almost all pharmaceutical, industrial and military chemicals. Other countries were anxious about the dominance of German companies. The Americans, playing catch-up, saw an opportunity. After Germany signed the Armistice in November 1918, the Allies confiscated the German dye industry's factories and properties – and, crucially, their patents. Now they had to decide what to do with them.

During the war, the US government had passed the Trading with the Enemy Act of 1917, which set up the Office of the Alien Property Custodian to control enemy-owned property. After the war, the Alien Property Custodian was J. P. Garvan, an industrialist who wanted to build a long-term future for American chemistry and medicine. He had a personal motivation too. The previous year, his eldest child Patricia had fallen ill with rheumatic fever, a complication that can arise from a simple bacterial throat infection. As a powerful politician, Garvan had access to the country's best doctors – but there was no treatment available, and Patricia died. Garvan was determined that a cure should be found. Salvarsan had shown that chemical medicine was possible, but the patent debacle had also shown how it could be controlled by unscrupulous interests.

As custodian of seized German property, Garvan wielded huge power. To start with, he wanted to ensure that German chemical patents weren't returned to German companies and their agents. First, he persuaded President Wilson to create a new public organisation by executive order: the Chemical Foundation had Garvan as its president and just five wealthy stockholders. Then, wearing his hat as the Alien Property Custodian, Garvan arranged for more than 4,500 German patents to be sold to the Chemical Foundation at a knockdown price of, on average, just $50 per patent. The patents covered a huge range: dyestuffs, explosives, medicines. Finally, as president of the Chemical Foundation, he began leasing the seized German patents out to American firms, making large profits which, after the stockholders got their cut, he could plough into medical research. The one patent that Garvan didn't sell off was Salvarsan: he left the licences for that in the hands of the US government. Metz was furious. He later said that to get the Salvarsan patent alone he personally would have paid $1 million.

Salvarsan was a beacon in the dark. In the following decades, German dye companies that had diversified into pharmaceutical research continued building on Ehrlich's programme, hoping to find the next Salvarsan for other infections. They tested new chemicals derived from dyes against microbes in test tubes and animals. At the factory of the German company Bayer in Elberfeld, one visitor in 1930 sang the praises of the site: it was 'the envy of the universities . . . erected by pure science: concentrated pharmacology!' The factory buildings that had once stood 'wrapped in steam' on green and violet ground, with the sharp scent of chemicals all around, had been joined by laboratories filled with *life*. Rooms buzzed with mosquitoes carrying malaria or squeaked with hordes of white mice; in one you could even hear the singing of unfortunate canaries. The scientist Wilhelm Roehl had worked with Ehrlich, and since 1922 had been in charge of chemotherapy at Bayer. Starting with methylene blue dye, the Bayer scientists found successive synthetic antimalarials that were even more effective. Roehl showed that they could be superior even to quinine. This was especially noteworthy, because quinine still depended on cinchona bark – the dream of synthesis had still not yet been achieved seventy years after William Perkin had attempted it.

But malaria was caused by a parasite. When it came to bacterial infections, Salvarsan remained alone, the sole realisation of Ehrlich's vision of bacterial chemotherapy. The streptococci, a group of bacteria named for the way their cells under the microscope resembled beaded chains (*streptos* meaning 'chain' and *kokkos* meaning 'berry' or 'seed'), were responsible for infections of the sort that had killed Garvan's daughter Patricia. There was still no effective treatment at the end of the 1920s – the author of one textbook stated that 'if one looks over the whole field of the streptococcal infections of humans, then one comes in every respect to an ever more resigned attitude'.

It was a streptococcal infection in March 1929 that earned Gerhard Domagk a sudden promotion at Bayer. Not his own: Wilhelm Roehl, his boss and the discoverer of the alternative to quinine, unexpectedly died at the age of forty-seven. Over the summer, Domagk had to assume temporary directorship of the chemotherapy lab. It was probably not a position to his liking, since he preferred experiments to administration. He had started to work at Bayer only on secondment from a nearby university, attracted by its resources.

Domagk was familiar with the horrors of bacterial infection. In WWI, he had served in a grenadier regiment in Russia, been wounded, and then transferred to the medical corps on the Western Front. Around half of the 10 million soldiers who died in WWI succumbed not to their wounds, but to the infections that followed. Of the thirty-three medical students who mobilised with him, thirty had died. The prospect of using Bayer's chemicals to treat disease, following in Ehrlich's footsteps, was a powerful motivation. Towards the end of 1932, Domagk was testing new chemicals concocted by Bayer's chemists against streptococci, the bacteria that had killed Roehl.

Domagk's superior, Heinrich Hörlein, was an advocate of Bayer's preferred approach to finding drugs: an ongoing and iterative screening process. The investment required was considerable. You began with a molecule you knew the effect of, and successively tweaked it to try to improve its performance, testing each new chemical and seeing if your changes had improved it. It was neither entirely rational nor random. In a way, it was almost an evolutionary process: the chemists were experimenting with dashes of randomness to see what worked, ending up with things that they could never have predicted from design alone. To target streptococci, the chemists were starting from a type of chemical known as azo dyes. The molecular structure of an azo dye resembled a pair of spectacles: each lens a benzene ring, connected by a bond called an azo bridge. Chemists would synthesise a variation of an azo dye, test it out in chemical reactions, then pass it to Domagk for his experiments.

To return to Ehrlich's lock-and-key metaphor, the chemists were like locksmiths, trying out slightly tweaked keys to open a closed door. At one remove, they had to try to intuit what was going on inside the lock. This meant dialogue with the medical researchers carrying out the tests against bacteria, sharing and discussing the patterns of results when they tried the compounds out. If better results seemed to come from dyes with modifications at a particular point on the compound – say, at a certain spot on the ring – the chemists could focus on further modifications at that site, using their toolbox of synthetic chemistry that had been steadily built up over the past seventy years.

Domagk had reported favourable results from one compound for streptococcal infections in mice: a burgundy-red dye. The chemists interpreted these results and prepared another similar chemical. The spectacles of the azo dye now had new side-chains of sulphur and nitrogen jutting off like adornments on novelty sunglasses. On 20 December 1932, Domagk reported his results with this molecule (KL370, later named Prontosil) to Hörlein. They were extraordinarily clear: all the mice treated with Prontosil were alive at the end of his experiment.

Hörlein knew the importance of patents. As he'd once put it, some other chemical companies were 'not all that different from the robber knights of the middle ages'. He believed in patents as a

justified way to recoup the investment in the large screening programmes needed to find magical substances. He moved quickly,
filing a patent on Christmas Day for a method of preparing new
azo compounds that killed bacteria. There were delicate issues of
timing and secrecy. If the German patent was granted, the information in it would eventually be published. German patent law only
extended as far as Germany's border; companies in neighbouring
countries would lose no time in making a chemical themselves if
they found out it was valuable. In early 1933, the first clinicians in
Germany were given Prontosil to test in patients.

As this work quietly continued, checking whether Domagk's
mouse experiments would translate to real-world efficacy, the obvious precaution was to also file a patent in France, still considered a
'country of counterfeiters'. However, Hörlein knew that as soon as
French companies became aware of the compound, they would start
manufacturing it. Domagk's work remained secret, with no publications outside the company. Gambling, Hörlein waited another
year to file a French patent on Christmas Eve 1933. Unfortunately,
the French system moved quickly and the patent was granted. The
result was that by June 1934, the secret of Prontosil had been published in France – six months before the German patent was even
granted. And unfortunately, French law offered less protection than
in Germany: patent protection didn't apply if the chemical was
used as a medicine. French companies easily intuited Prontosil's
medical promise and commenced their own investigations at speed.
Domagk's first scientific paper announcing the results of his experiments with KL370 was published in February 1935, but the French
had made good use of their head start. By May, a French company
was already selling Prontosil under a different brand name.

In Germany, Prontosil was successful as soon as it went on sale in
1935. That December, almost three years on from the dramatic
results of Domagk's first experiment on mice, should have been a
happy time. The Domagk family were preparing Christmas decorations. Domagk's six-year-old daughter Hildegard was coming down

the stairs of their house with a needle when she tripped, falling and driving the needle deep into her wrist. At first, it didn't seem too serious – Domagk took Hildegard to have an X-ray and the needle was removed half an hour later. The following day she had a fever, but Domagk tried to reassure himself that this was merely a lasting symptom from a sore throat she'd had a few days earlier. But over the next few days the fever developed, and the wound site swelled. The needle had created a dangerous inroad into her body and an infection was growing, spreading up Hildegard's arm. Her condition was deteriorating day to day. Domagk tried to remain practical. He took a bacterial culture and identified the culprit: the dreaded streptococci.

Of all the parents who had been in that alarming situation before him, he was the first who had Prontosil. Four days after the accident, as Hildegard lay dizzy and feverish on her bed he gave her the medicine he had helped discover. So great was his concern for his daughter that he gave it by both means available to him: as an injection and in tablets which he pressed gently into her mouth. Once inside her bloodstream the molecules of Prontosil began to flood around her body, magic bullets to target the bacteria that were killing her.

Hildegard's life was saved, but Domagk was quite unaware of the transformation his wonder drug had undergone in her body. Enzymes in her liver, as well as the resident bacteria in her gut microbiome, deconstructed Prontosil: they snapped the azo bridge, breaking the molecular spectacles in two. The active part of Prontosil was just one half of the spectacles – a simpler molecule known as sulfanilamide. Ironically, the synthetic therapy created by Bayer's chemists relied on the collaboration of life itself to forge the bullets. This bizarre twist in the tale was quite unknown at the time, but would come to be recognised as a general phenomenon of a 'prodrug'; Prontosil was a substance that was transformed inside the body into its active form.

Just weeks before Hildegard tripped, French scientists had discovered through a different route that the simpler molecule, sulfanilamide, was really responsible for Prontosil's effects. In November

1935, they had started testing modified versions of the Prontosil molecule on mice, hoping to find improved medicines. Each different molecule was a reconfigured pair of the Prontosil spectacles, made from combining a colourless intermediate compound – sulfanilamide – with a different-coloured lens. The scientists were treating mice in allocated groups, but ran out of the coloured compounds when it came to the final group: they had four mice left over. On a whim, one researcher decided to give the four mice some of the sulfanilamide they had been using to prepare the other molecules. When they came back the next morning, those mice had all survived. At first, the French scientists were disheartened – all that time and effort preparing the various complicated dyes had been 'completely useless'. But then they realised the significance: all of Prontosil's power came from just one half of the molecule. The medicine's beautiful burgundy-red was superfluous. They wrote up their findings and published them at the end of the month.

At first, the German chemists didn't trust the claims from their French rivals. Prontosil was a dye firmly in the proud tradition of Ehrlich, and it's possible they simply found it hard to believe a *colourless* compound was capable of such effects. One urgent question was what this meant for the patent situation. It turned out that the colourless sulfanilamide had been patented before. In 1909, chemists at Bayer had patented the new compound, extracted from coal tar by an Austrian chemist, as an excellent intermediate for making new azo dyes for wool. That patent had gone on its own journey: after WWI, it was seized in the American grab-bag of 4,500 patents; when these were returned, it had passed in 1925 after the giant merger of German dye companies into the property of a consortium, IG Farben, that included Bayer. And then, in 1929, its twenty-year period of protection under German law had expired – six years before Bayer was about to launch Prontosil, which was now revealed as simply a coloured version of the real active ingredient. Even within Germany itself, sulfanilamide could be manufactured by anyone, free of charge.

Hörlein, the man who had been so careful with managing the

Prontosil patents, must have felt more than a flash of annoyance. The 1909 sulfanilamide patent had been filed by the responsible person at Bayer at the time: himself.

From 1935, it became widely known that sulfanilamide was not itself patentable, but hundreds of related derivatives were – each a separate molecule, but with the same 'sulfa' prefix. The development of these sulfa drugs proceeded like a gold rush. But as the sulfa drugs were booming, Europe seemed to be inching towards yet another catastrophic war. The chemical and pharmaceutical industries in Germany that had brought about so many medical advances became increasingly entwined with the Nazi state. Germany was transforming its economy. To many, conflict seemed inevitable.

At the same time, the war against bacteria was far from won. Chemists had a lot to learn. The human body wasn't just an inert receptacle for their products to work their magic in; it was itself a chemical factory, full of its own intricacies. That perspective also applied to bacteria; rather than just being targets for a magic bullet, they were active in their own right. Prontosil's life-saving properties relied on a final enzymatic twist – courtesy of bacteria – that cleaved the sophisticated molecule the chemists had constructed.

In *A Study in Scarlet*, the first of the Sherlock Holmes stories that Paul Ehrlich loved, the investigation is nearly thrown off by a false clue. A man is found dead with a word written above him in blood: *RACHE* (the German for 'revenge'). The clue baffles Scotland Yard, particularly when it appears again on the wall above a second murder. But Holmes concludes from the handwriting that it was not written by a German, but by 'a clumsy imitator who overdid his part'. In the end, the murderer admits that he simply wrote it as 'some mischievous idea of setting the police upon a wrong track'.

Sherlock Holmes isn't fooled by appearances. He wants to follow his own red thread: the 'scarlet thread of murder running through the colourless skein of life', to 'expose every inch of it'. Ehrlich's red thread – the chemistry of dyes – had led scientists from the black sludge of ancient forests to the most effective antibacterial drugs ever discovered. From the blackness of coal tar, formed over millions of years, had come Perkin's first purple dye. The notion of a magic bullet had driven researchers as they worked out how to manipulate molecules in the hope of targeting the unseen locks of microbes. Salvarsan had proved that it was possible, and Prontosil had delivered even further on that dream: a dye that saved lives. Dyes had been essential for the foundation of bacterial chemotherapy – but it turned out that for all their beauty and vitality, their colours had been an enticing distraction, irrelevant to their antibacterial activity. The chemists at Bayer had, as historian John Lesch puts it, been entranced by the 'mystique of colour'.

Still, the thread was real. The risk in history is that knowing the end of the story makes certain events look like wrong turns. But Salvarsan, Prontosil and the other dye-based antibiotics were a spectacular achievement. They marked the transition into a world where cures for bacterial infection became a reality. In just a few decades at the start of the twentieth century, the whole concept of what medicine could achieve had been radically altered.

Across human history, most medicines had been derived from plants. The forest had been the source of medicine. Now, science could summon medicines out of ancient forests, forging new molecules from ancient hydrocarbons in combination with other chemicals. The fights about ownership were now not over patches of rainforest, but over intellectual property. Governments and companies tackled each other inside an evolving system of laws that had never been intended to apply to medicines. Control of new industries was now central to global power. The booming pharmaceutical industry had emerged from the dye industry, which in turn had emerged from fossil fuels. Now, for better or worse, that pharmaceutical industry was synonymous with medical progress.

Prontosil meant that the infections that had suddenly killed Garvan's daughter Patricia and cut Wilhelm Roehl's life short were now treatable – but, curiously, not always. It wasn't entirely explicable: some patients' infections melted away in the face of treatment with sulfa drugs, while others relapsed and died. By 1940, at least one paper had speculated that resistant strains might be responsible. Ehrlich had noticed the phenomenon of resistance as early as 1907 but was unconcerned: he saw resistance as a tool for studying cellular functions. Antibiotic resistance was not, at first, perceived as a major problem. Rather it was a curiosity.

In 1940, one year into WWII, sulfa drugs had become indispensable for treating streptococcal infections. They may even have changed the course of the war: a sulfa drug made by a company called May & Baker was used to treat the British Prime Minister Winston Churchill's bacterial pneumonia – twice. In a personal communique, he said that the 'admirable M&B' had 'repulsed' the bacterial intruders from his body. But by the end of the war, an even more remarkable antibiotic would become widely available. The chemists were about to receive a challenge to their supremacy. The new antibiotic didn't come from the remnants of long-dead forests; it was made by a microorganism.

More lives than war can spend
Penicillin

There is something . . . to be learnt about dust, the dust of every-day life.

The Irish Homestead

In the middle of 1944, Nazi scientists claimed they were close to perfecting a wonder drug. Research into making this new antibiotic had been underway for about six months, led by Hitler's personal doctor, Theodor Morell. Hitler had taken a keen interest in the research, using his power and influence to ensure that no expense was spared. He had personally bought Morell an advanced microscope, exempted his company from tax and awarded him the highest civilian honour of the Nazi regime. Morell claimed that although he had personally discovered the new antibiotic years ago, the British Secret Service had stolen the discovery and given the credit to an obscure Scottish doctor – a man called Alexander Fleming.

Morell was a fraud. The powder he sprinkled on the Führer and called penicillin – for that was what he claimed to have invented – was bogus and impure. Yet the true story of penicillin was, if anything, even more incredible. Just four years previously, the world's entire supply could have fitted on a teaspoon. But following the outbreak of WWII, thousands of scientists had desperately worked to scale up production. They had transformed an initial observation of impure mould juice into the world's most valuable medicine. That mould had been grown in hundreds of biscuit tins in Oxford, transported to America in a suitcase, and grown in huge industrial vats in the plains of the Midwest. Now, in June 1944, it made a triumphant return to Europe, carried in hundreds of thousands of vials by the Allied soldiers who cascaded onto the beaches of northern France.

The era of penicillin had begun. It had all started with a handful of dust.

Wherever you are reading this, you are almost certainly no more than a few metres away from a member of *Penicillium*. They are a diverse group of fungi – members of the eukaryotes that can be either single-celled or multicellular – containing hundreds of different species, found all over the world and almost everywhere. *Penicillium* can be found in dust, in water, inside the walls of buildings and on rotting food. As you read this, some *Penicillium* are hard at work turning milk into cheese in French caves (*P. roqueforti* for roquefort, *P. camemberti* for camembert). Another, *P. digitatum*, will be blooming its green fuzz on an old orange on a Japanese table. Still another, *P. lacus-sarmientei*, is lurking in sandy soil by a remote lake in Tierra del Fuego. Most are harmless, but like other fungi they can exacerbate conditions like asthma and even be deadly for people with weakened immune systems. When the writer Bruce Chatwin contracted a serious fungal infection with a fungus known as *P. marneffei*, he claimed that doctors had told him it had previously only been recorded in ten Chinese peasants and a dead killer whale.

What all *Penicillium* members have in common is their botanical appearance. In 1809, the first recorded scientific description of *Penicillium* talked of it as a plant: its 'grassy tufts' had 'branches with a brush tip' (*penicillus* means 'brush' in Latin). Under the microscope, *Penicillium* bodies looked like fistfuls of berries at the end of slender branches. Even well into the twentieth century, the place of fungi in the great tree of life remained enigmatic and mysterious. It would take genetic analysis to confirm that, though they look like plants, they belong within their own distinct kingdom.

Fungi, like us, have had to learn to live in a bacterial world. Dealing with that challenge produced two forks in the road: most animal cells went one way, fungi another. Filamentous fungi like *Penicillium* spread out over environments in search of food with long slender cells called hyphae. These fungi weave their hyphae into a mesh-like network called a mycelium. One ancestor of *Penicillium* that lived 400 million years ago wove mycelium-based structures almost a metre in diameter that resemble tree trunks – so much so that their fossils were at first mistaken for conifers. The *Penicillium* genus itself branched off from its nearest fungal relatives much later, when dinosaurs walked the earth in the upper Cretaceous period between 60 and 85 million years ago.

Filaments are only one means of fungal propagation. Fungi can also use tiny spores to spread further afield – the equivalent of seeds, light enough to float away through the air. Some fungal species release spores altogether in a plume that reduces drag, forming the equivalent of a peloton of cyclists that makes them travel twenty times further than they would otherwise. Each year, an estimated 50 million tonnes of fungal spores enter the air. About 10 per cent of the fine particulate matter in the outdoor air is made up of spores. *Penicillium* species are the most frequently detected indoor fungi, their love of damp and decay making them one of the few beneficiaries of old, cold buildings. Their spores are consequently among the things we breathe in the most: each spore is around five microns in diameter, meaning you could line up one hundred across the full stop that ends this sentence. An invisible dust of *Penicillium* floats inside most buildings. Outside, these spores can be caught by the wind and lifted to great heights; they have been recovered from an altitude of 20,000 metres.

In 1928, Alexander Fleming was working as a bacteriologist in St Mary's Hospital in London, when he noticed an instance of contamination. He'd been growing bacteria known as staphylococci: under the microscope, their cells cluster in bunches like fat grapes. But a *Penicillium* spore had landed on one of his Petri dishes. The *Penicillium* had grown too, and produced a circle of exclusion around it: a demilitarised zone where the bacteria could not grow. Fleming suspected that something in the 'mould juice' was killing the staphylococci. He named that filtered mould juice 'penicillin'.

Given that *Penicillium* spores are so ubiquitous, it may seem hopeless to imagine that we could trace the passage of the *Penicillium* spore that danced on eddies of air before it landed on Fleming's Petri dish. But not all *Penicillium* species produce penicillin. (For example, though camembert is a wonderful substance, it contains no trace of penicillin because *P. camemberti* doesn't make the molecule.) So where did Fleming's spore come from? The apocryphal story is that the fungal spore must have entered through the open window, but it's almost certainly wrong: Fleming's window opened onto a busy London road and was usually firmly shut. What's more, an Irish scientist called C. J. La Touche had a laboratory on the floor below – a room stuffed to the rafters with moulds.

In the 1920s, a prominent theory held that asthma was worsened by exposure to mould. To investigate whether asthma patients could be desensitised to the specific fungi in their own home, La Touche had collected dust and floor sweepings from the homes of asthma sufferers. From these, he cultured fungi and identified them. So, when Fleming wanted to identify his contaminant, he naturally sought the advice of his colleague downstairs. La Touche identified Fleming's strain as *P. rubrum* (its name meaning 'red'). This was a species that one French scientist had called an *animal de laboratoire* – one that lived not outside in nature, but inside buildings. Fleming found that the strain had antibacterial activity against a huge range of bacterial species, although Gram-negative organisms seemed to be resistant to it. Fleming was intrigued to see if other fungi could also produce a mould juice with similar properties and experimented with the other strains in La Touche's large collection. He could find only one, but the broth it produced was eerily identical in its properties. Though some say Fleming's *Penicillium* drifted in through an open window, it seems much more plausible that spores from La Touche's mould had floated up to Fleming's laboratory from the floor below.

Whether or not Fleming realised that La Touche's lab was the probable source of the *Penicillium* contamination, his failure to reproduce his observations in other *Penicillium* species showed that he'd been lucky. The degree of coincidence involved was not just the presence of the right spore in the right place, but also a set of additional circumstances: the spore needed to land on a Petri dish at the right time with the right temperature and the right conditions. If there hadn't been stacks of dirty Petri dishes waiting to be tidied away, it might not have happened. But though Fleming was not the first to happen upon the antibacterial properties of *Penicillium* – previous scientists had recorded them before him – he proved the most far-sighted when it came to the possible applications of that chance discovery.

However, though Fleming was able to reproduce his experiments and eagerly sent his precious strain out to the few people who asked for it, nobody had much luck developing anything further. Try as they might, they couldn't isolate the active ingredient. And without that, there were no prospects for medicine. Fleming made some lacklustre attempts: he cultured the strain in milk to make a sort of penicillin cheese that his obliging lab assistant ate ('very much like Stilton'); he wrapped an amputee's stump with a clammy dressing soaked in mould juice. The resulting 1929 article he published offered a single practical use for his *Penicillium* strain: preparing contamination-free cultures of *Bacillus influenzae*, a species immune to the effects of the mould juice. His mould juice wasn't only an unappealing medicine – it was impractical and possibly dangerous. *Penicillium* strains can make molecules that are toxic for humans too, known as mycotoxins.

To any modern doctor schooled in disinfectant and antiseptic technique, mould juice had a pungent whiff of medieval folk remedies: slapping poultices of mouldy bread on cuts, filling abscesses with soil or rubbing yeast onto boils. If these were known about at all, they were seen as primitive superstitions, based on a belief that dirt could cure infection by some strange sympathetic magic.

Penicillium strains that made penicillin, a medicine which in a little over a decade would be valued weight-for-weight more than gold, were everywhere. They lurked in everyday life: sometimes in the blueish bloom on bread, sometimes in damp and dark corners, yet only ever producing penicillin in the most infinitesimal quantities. Penicillin was right in front of Fleming, floating in his impure mould juice, but he couldn't pluck the chemical out. His observations were a little like noticing that a lump of ore shone prettily, without having any idea of the processes needed to separate out its gold.

On its publication, there was no indication that Fleming's dry and technical paper contained the germ of an idea that would transform medicine. Indeed, from 1935 the introduction of Prontosil, followed by the other sulfa drugs, seemed to leave his flight of fancy in the dust. These were medicines that could be made reliably and predictably, millions of times over, unlike Fleming's furry and sticky attempts. But the massive success of the sulfa drugs had an important side effect: it prompted other scientists to look back through old publications, searching for other chemicals that might be able to serve as medicines. At Oxford's Dunn School of Pathology, a Jewish émigré from Germany called Ernst Chain found Fleming's article. Chain resembled a neater version of Albert Einstein; like Einstein, he was a talented musician, and had considered becoming a concert pianist. He thought Fleming's mould juice might be of interest, if the active chemical responsible could be isolated: the gold extracted from the ore. The head of the Dunn School, the Australian scientist Howard Florey, agreed. It turned out they didn't need to write to Fleming to get the mould juice. Chain discovered that the Dunn School already had a culture of Fleming's strain, obtained years before and still used, as Fleming had suggested, for the mundane process of preparing pure cultures of *Bacillus influenzae*.

The Oxford scientists soon became excited by the *Penicillium*. If they could isolate its juice's hypothetical active ingredient, which was what they called 'penicillin' (where Fleming had used the word for the juice mixture as a whole), they might have a drug to rival Prontosil. Chain showed he could extract the penicillin, although its purity was difficult to assess. Penicillin's extraordinary power can be seen in the original unit of measurement the Oxford team defined. A unit measure of penicillin was defined in a roundabout way as the amount that needed to be dissolved in fifty millilitres of meat extract to stop a standard strain of staphylococci from growing. That was only 0.6 micrograms of penicillin – around a thousandth of a grain of salt. But even getting such a tiny amount was problematic.

Their experiments had shown that the *Penicillium* grew best on the surface of a liquid broth. The ideal depth of liquid necessitated a broad and shallow vessel, rather than the standard laboratory glassware of test tubes and flasks. In 1940, Chain handed over the task of growing Fleming's strain to a British biochemist called Norman Heatley, a keen twenty-nine-year-old who was adept at making his own equipment. Heatley inherited the vessels Chain had been using to grow *Penicillium*: two pie dishes. He quickly abandoned them and scouted around for better alternatives, settling on biscuit tins – hundreds of them.

Heatley was a master of improvisation. At a time of wartime shortages, this was essential. He had already used whatever he could find to build a machine for separating out the precious penicillin itself from the mould juice. Heatley's device consisted of conventional laboratory equipment mixed with a jumble of parts that could politely be described as pre-loved: an old bookcase scavenged from Oxford's Bodleian Library, an old doorbell, coloured warning lights and copper cooling coils. This was science on a shoestring budget – although shoestrings were one of the few things Heatley didn't use. Heatley's improvisation paid dividends. By March 1940, he'd cobbled together one hundred milligrams of penicillin. But they needed more.

As Heatley tried to procure biscuit tins, there were dramatic developments in the war. On 10 May 1940, the lightning-quick German invasion of France stunned the world. British soldiers who had been stationed in northern France began a desperate retreat, culminating in the evacuation from the beaches of Dunkirk which began on 26 May. On the same day, the Oxford scientists got their first proof that tiny amounts of the extracted penicillin could rescue mice from otherwise fatal bacterial infections. The race was on to scale up production. Just over the English Channel, Hitler was in Paris by the end of June, touring the Eiffel Tower and the Arc de Triomphe – all now under German control.

Not for the first time, the progress of war was outpacing medicine.

In war, people kill people, and bacteria take care of the rest. Any weapon will do, however old-fashioned: swords, arrows and bullets all break the skin, allowing bacteria to flood unimpeded into open wounds. But modern warfare had led to terrible forms of progress. Fleming had witnessed these developments first-hand. At the start of WWII, he was nearly sixty, but he had seen what modern technology was capable of as a young doctor in the trenches of WWI. From the second half of the nineteenth century, the dye industry that had given the world new medicines had also led to new chemical explosives: dynamite, gelignite and TNT. These new destructive materials were used to create powerful artillery weapons that blasted fragments of *anything* deep within shattered bodies: metal, dirt or clothing. All could be contaminated with bacteria. Working as a doctor in northern France, Fleming showed that gangrene-causing bacteria lurked in nearly all soldiers' uniforms. In artillery explosions, these bacteria were given sudden and gory access to soldier's bodies where they could feast in abundance, rotting and blackening the flesh, before killing in an excess of putrefaction and sepsis. WWII promised much the same, unless better antibiotics could be developed.

As the situation in Europe worsened, the Oxford scientists talked candidly about the possibility of a German invasion of Britain. If they needed to flee, they could hardly gather up a roomful of biscuit tins slopping with broth. Heatley suggested that they put their trust in the *Penicillium* spores: he and his colleagues rubbed a pinch into the linings of their coats. Even if they had to leave with only the clothes they stood up in, they would be taking Fleming's strain with them. But as Heatley had discovered, Fleming's strain was proving inconsistent. The issue was not simply having the strain, but coaxing it into doing what the scientists wanted.

Inside the Dunn School, Fleming's strain was being grown at desperate scale, as blue-green mats of fungal mycelium floated in the biscuit tins. But Heatley had discovered that the strain's penicillin production was variable; some spores never became penicillin-producers because of some unknown natural variation. He chose carefully based on those spores that gave the highest penicillin yields and kept culturing from their progeny, like a pigeon fancier breeding only from his favourite birds. In *On the Origin of Species*, Darwin had noted that 'the key is man's power of accumulative selection: nature gives successive variations; man adds them up in certain directions useful to him . . . The great power of this principle of selection is not hypothetical.'

Heatley had to deal with the fact that a population is not a fixed quantity, but a repository of possibilities. Imagine the potential variation of *Penicillium* as a series of valleys and hills, where height represents how successful an organism will be. At every generation, survival is dependent on altitude. Over time, the valleys will be depopulated, and organisms will end up clustered at the highest points. This metaphor captures the principle of natural selection that Darwin described, often summarised as the survival of the fittest. (Here 'fittest' shouldn't be taken to mean the strongest, but rather the sense of being *fitted* to an environment.)

Now imagine a particular microbe. Natural selection will mean that it should sit at the top of a hill – not, perhaps, the highest hill for miles around, but a comfortable lookout over evolutionary possibility. But a microbe is never just one thing: every possible single mutation in its genome – of which there will be millions – is like a tiny step on the landscape. Viewed over time, a microbial population is a collective unconscious, shimmering with random flutters around a central archetype. Just as physicists view electrons not as single point-like particles but as clouds of probability distribution, so biologists view populations as collections of variation. Natural selection keeps the population tightly bunched at the fittest genome at the hill's summit. The mutants are like millions of medallists jostling around a tiny central podium, with most of them flickering in and out of existence in a few generations – from the top, every direction is downhill.

But the landscape is not fixed and eternal. A change in conditions can come like a tectonic shift where the whole landscape is reconfigured: sinkholes open on the slopes, a high peak may collapse into a deep pit, a valley floor may ascend into the sky. Those microbes who were at the edge of the cloud, dwindling away, may find themselves the best hope for survival, higher on the slopes of the brave new world. If they prosper, they in turn become the new archetype.

Why did the *Penicillium* sometimes stop making penicillin? In nature, penicillin presumably provided it with an advantage. But it took metabolic energy to make. It's possible that as Fleming's strain was grown again and again in the safe and bacteria-free environment of the Oxford labs, penicillin was jettisoned by some members of the population because of mutations which reduced or shut off production. By intervening and choosing only the best penicillin producers, Heatley was exercising the power of selection, shifting Fleming's strain little by little towards what he wanted. From his initial experiments, by Christmas 1940 he had scaled up penicillin production around a thousand times, but the yields were still far too low to use as a medicine.

At the same time, his colleagues were discovering that evolution cut both ways. At the end of 1940, Chain and Edward Abraham, another Oxford scientist, reported that they had discovered an enzyme which could destroy penicillin, found in the common bacterium *E. coli*. This was an early hint that for any antibiotic, resistance probably existed somewhere out there in the uncatalogued diversity of bacteria. The churn of biochemistry and the near endless variation of microbial life meant that anything evolution could make, it could also unmake. At this point, penicillin had never been used in patients, but already there were signs that its evolutionary counterpart was waiting in the wings. In evolution, there is no perfect strategy that will guarantee eternal victory. The landscape is always shifting; the only constant is change.

By the start of 1941, the Oxford team had already proved penicillin's effectiveness in animal tests. They had now obtained enough to consider human use. Despite Heatley's ingenuity, obtaining a daily dose for a patient (about sixty milligrams) meant brewing about seventy litres of raw *Penicillium* broth. That meant harvesting about fifty dishes a day, each of which took at least a week to brew. It was an inefficient nightmare, so they had to choose patients carefully.

Their first patient in Oxford didn't have an infection. Elva Akers was a fifty-year-old woman with terminal breast cancer. She bravely volunteered to receive the first injection of penicillin: a doctor simply asked her if she was willing, and she agreed. After the injection, she suffered violent side-effects, becoming feverish and stiff. These horrible effects would surely lead to a suspension of any medical trial of an untested drug today. But the Oxford team weren't going to stop. They blamed the reaction on impurities in the preparation of the penicillin (probably correctly) and continued as before.

On 12 February, penicillin was used for the first time on someone who truly needed it. Albert Alexander was a policeman who had been injured by shrapnel in a German bombing raid. His facial wounds had become infected – he'd already lost one eye – and the bacteria had spread into his bloodstream. He was in a desperate state; Heatley wrote in his diary that Alexander was 'oozing pus everywhere'. The treatment used up the entire global supply of penicillin almost immediately. They knew that about half of the precious supply was literally being pissed away – because penicillin isn't metabolised, it was being passed out in Alexander's urine. Researchers collected his urine, then transported it by bicycle back across Oxford to the Dunn School laboratory where they extracted what penicillin they could (they called it the 'P-patrol'). Unfortunately, it was a law of diminishing returns. As the same penicillin cycled through his body again and again, a smaller and smaller amount could be extracted each time. Yet Alexander began to recover: five days of treatment produced dramatic improvement. Unfortunately, after the treatment stopped, despite ten days of stability, he then relapsed into fever and eventually died on 15 March.

The Oxford scientists had been given a tantalising glimpse of what was possible. Treating Alexander had been a huge effort in a country that was struggling to provide its citizens with the basic necessities of life; mass penicillin production was unimaginable. Just before Alexander's treatment began, Churchill had made an impassioned radio address directly to the United States, still at this point officially neutral, to supply the British economy with what it needed to keep fighting the war: 'Put your confidence in us,' Churchill urged. 'Give us the tools and we will finish the job.' Just four days before Albert Alexander died, President Roosevelt had signed the Lend-Lease Act into law. The United States was now committed to supply Britain and other Allied nations with billions of dollars' worth of materials: everything from food to industrial chemicals to military trucks. Slowly but surely, America was being drawn into the war.

In this wider geopolitical context, spores from Fleming's strain became airborne once again, soaring over the Atlantic in a Boeing 314 Clipper seaplane to arrive in a humid New York on the afternoon of 2 July 1941. Penicillin production was about to be given the full support of the American economy.

Peoria is a quiet farming city in the heart of the American Midwest, lying on the broad Illinois River. It was the unlikely location of the next transformation in penicillin production. A group of federal microbiologists known as the 'mould kings' had recently been redeployed there from Washington, DC, and Heatley was going to work with them on *Penicillium*.

The first step was to germinate the spores that had travelled from Oxford. At first it seemed they were dead, but after a tense few days they started to grow; they were probably unaccustomed to the relative heat of the Midwest summer. The next step was to improve the penicillin yield. Back in Oxford, Heatley had experimented with various concoctions for the nutrient broth he fed the *Penicillium* – cow muscle, glucose, even the British yeast spread Marmite. The Peoria laboratory were adept at using agricultural by-products, and in America's expanding economy, everything was meant to be useful. Now, they tried out their own secret weapon: corn steep liquor.

Corn steep liquor is a brown and sticky concentrate that reeks of booze gone bad, a waste product of breaking corn kernels down by soaking them for days in water. It was so undesirable that it was often dumped straight into the Illinois River that flowed through Peoria. At sixty cents a gallon, it was a cheap food for microbes. Adding the liquor to the *Penicillium* broth increased its penicillin yield, and within the year, corn steep liquor had been designated an essential wartime product that was subject to official price controls. Adding another agricultural product, milk powder, improved yields even further. Compared to what had been managed in Oxford, the same mould was now generating over fifty times more penicillin. Back in Oxford, the bodged extraction apparatus continued to expand in Heatley's absence. It now included milk churns, a steel bathtub, a bronze letterbox and a couple of aquarium pumps. It would remain the largest extraction plant in Britain for another two years. The contrast with the industrial production being fired up in America was pitiful.

Fleming's strain had had a long journey from the London dust. Its ancestors had needed penicillin to fight over sparse nutrients with other microbes in the damp and the cold; now it faced the prospect of being given all the food it could ever eat, so it could be devoted to the production of penicillin. But rather than simply working with Fleming's strain, the Americans had started a search for superior penicillin producers. After they'd joined the war in December 1941, all US Army Transport Command planes were sending back soil samples from wherever they landed, which were then screened for penicillin-producing strains. But the scientists in Peoria also knew that *Penicillium* lurked in decaying fruit, so one laboratory assistant was given the unenviable job of finding samples – her colleagues dubbed her 'Moldy Mary'. Amazingly, a rotting cantaloupe melon contained a *Penicillium* that made several times more penicillin than Fleming's original strain.

In a year, the Americans had tweaked and selected their way up to more than a hundred times greater yield. Yet there still wasn't enough penicillin for more than a handful of lucky individuals. Given the success of the sulfa drugs, it was logical to think that the key would be synthesising penicillin chemically – just as had been hoped for quinine in the nineteenth century. Over a thousand chemists were trying, but they didn't know the drug's structure, and they seemed to be getting nowhere. Given enough time, the microbiologists could keep curating the natural variation in the *Penicillium* populations and gradually increase the yield of penicillin. After all, evolution is the product of variation and time under selection. But there was no time.

So, they tried the nuclear option: literally. Research on the effects of nuclear radiation was at an early stage, but scientists knew that it could induce violent and sudden changes in organisms, from deadly cancers to the breakdown of the whole body. That meant an increase in the variation available. The mechanism for these changes wasn't known at the time, because DNA hadn't been conclusively identified as the carrier of genetic information. Beyond the purple hues of visible light in the spectrum lie more energetic forms of light: ultraviolet radiation and X-rays. Nuclear radiation produces huge amounts of concentrated energy in the form of photons with these high energies, able to penetrate through matter. But sometimes, if one of these energetic photons passes through the cell and hits a molecule of DNA, it can physically damage the double helix like a cannonball careering through a ladder. Such radiation can have multiple effects: it can split apart the rungs (a 'double-strand break'), induce a mutation at a single letter of DNA, or generally produce chaos as the mechanisms that keep DNA in check go awry under the onslaught. These sorts of random changes are the raw material of evolution, but only under intense nuclear radiation do they accumulate so quickly. Most of the time these mutations are deadly, but sometimes they can produce a vast and beneficial change – such as an increase in production of a molecule like penicillin.

A Yugoslav-American scientist called Milislav Demerec and his team in Long Island were given the *Penicillium* strain to irradiate. Colleagues in Wisconsin followed up with further bombardment. If the *Penicillium* could scream, they would have heard it. Huge gobbets of energy slammed into its DNA, making the genome mutate at high speed. Out of 5,000 mutated strains, they discovered a new strain with a substantially higher yield of penicillin. (A 2014 paper sequenced the DNA of one of these mutant strains and found it had two copies of the cluster of genes that made penicillin: exactly the sort of random change that can be produced by a frenetic hail of radiation.)

As the war continued, scientists in other countries also attempted to make penicillin. In occupied France, scientists secretly worked with pre-war strains, worried that the Nazis would discover their work. In China, scientists looked for penicillin-producing strains on bean curd and shoe polish. In Russia, penicillin was produced from *Penicillium* scraped from an air-raid shelter. In Germany, scientists believed they'd got their hands on Fleming's strain from French scientists, but made little progress with it – it's possible they were deliberately sent a dud. Across all these scattered efforts, no other country came anywhere close to the scale of American penicillin production.

The American production of penicillin grew rapidly, fuelled by the world's largest economy. American efforts were not only better financed but better organised, as the US Committee on Medical Research enlisted multiple companies as collaborators in producing penicillin. These were sometimes unexpected: in the middle of 1943, most of the world's penicillin was being produced by a company called Chester County Mushroom Laboratories in Philadelphia, which usually grew mushrooms for food and had no prior experience in drug development. Participants in the penicillin programme were encouraged to share technological expertise across companies and universities. One revelation was a switch in fermentation methods: rather than having to maintain small sloppy pans of mould and scrape the scummy product off the top, it was possible to grow *Penicillium* in a deep rotating cylindrical drum.

Pfizer, at the time a chemical supplier rather than a pharmaceutical company, was a pioneer in such 'deep-tank' fermentation for producing citric acid. In 1943, it built the world's first dedicated penicillin plant, using spare parts including a used elevator and a second-hand boiler.

Others copied Pfizer's fermentation methods, and by the end of the war, over twenty companies had participated in the government programme for penicillin production. The emphasis on collaboration accelerated technological progress, with usual rules about competition and capitalism suspended. As one US government poster put it, this was 'a race against death'.

That wasn't entirely true. Penicillin wasn't only a life-saving medicine. It had also proved enormously useful for treating gonorrhoea, a sexually transmitted bacterial infection. In North Africa in May 1943, soldiers who were preparing to lead the Allied invasion of Sicily had caught sulfa-resistant gonorrhoea from brothels in cities like Tunis and Algiers. The only other treatment was inserting a catheter into the penis and irrigating the urethra with stinging chemicals – and it didn't really work. In contrast, a single day's treatment with penicillin was 98 per cent effective at returning a soldier to fighting fitness. The British soldier in charge of penicillin distribution at the War Office was Major-General Poole. His advisors debated using the limited penicillin available to treat those laid low by germs rather than Germans. Florey, the head of the Dunn School, who was travelling to North Africa to test penicillin for new conditions, was opposed, because they wouldn't learn anything from treating gonorrhoea – they knew it worked already. Poole's advisors were more concerned about the political fallout. What if the press found out? They imagined bishops in the House of Lords standing up to read out letters about soldiers who'd been wounded in the line of duty who hadn't received penicillin, while 'scallywags' had been cured. But a scribbled note in green ink, reportedly from Churchill himself, made clear that the scarce penicillin should be used for gonorrhoea: 'This valuable drug must on no account be wasted. It must be used to the best military advantage.'

Judged by the price the US government would pay for penicillin when this debate was happening, it was worth 250 times as much as gold. By 1944, there was enough to take it onto the battlefields to treat wounded soldiers as well. By the end of the war in 1945, the cost of American penicillin had dropped by a factor of 300, making it finally cheaper than gold. Four years previously, it had been the rarest medicine in the world. Now it would soon be available in almost every hospital in Britain and America – and therein lay a problem.

As WWII ended, the gold dust of penicillin was a powerful symbol of the hopes for a new post-war era. A triumphalist attitude prevailed: humanity had a medicine that killed bacteria while being almost totally non-toxic to humans. As early as 1945, the Nobel Committee was in no doubt that penicillin warranted the Prize for Physiology or Medicine. Although the creation of penicillin as a medicine had involved thousands of people, under Nobel's rules only three could share the prize. After intense lobbying, they decided that the glory should be shared equally between Fleming, Florey and Chain.

Fleming used the occasion of his Nobel Lecture in Stockholm to reflect on this triumph, but also to give a warning. Looking out at his audience, he spoke about the possibility of antibiotic resistance. He told a story about an imaginary patient, 'Mr. X', who took penicillin to treat a sore throat. In Fleming's framing, Mr X used enough penicillin to 'educate' the bacteria inside him to be resistant, but not enough to kill them. These 'educated' bacteria were then transmitted to his wife, who took penicillin but to no avail: she died. Fleming attributed responsibility for her death to Mr X, 'whose negligent use of penicillin changed the nature of the microbe'. Fleming summarised the moral of his story: 'If you use penicillin, use enough!'

Fleming's argument is often quoted today as a prescient warning, but it also shows something worrying: he misunderstood how evolution works. Fleming seemed to imply that resistance to penicillin could be avoided if everyone simply used enough of it – but this was wrong.

It would take another British scientist to expose his dangerous misunderstanding. Mary Barber, like Fleming, was a medical doctor who had moved into bacteriology. She worked three miles west of Fleming's laboratory in St Mary's, in a laboratory at Hammersmith Hospital, another elegant Victorian brick building. She was an expert in staphylococci, the same bacteria Fleming had first observed penicillin's activity against. Staphylococcal infections were vulnerable to penicillin and doctors reported wonderful results. However, it was clear that the proportion that were highly resistant to penicillin was rapidly increasing. Barber's own data showed that around 10 per cent of staphylococci seemed to be naturally resistant to penicillin. By 1947, two years after Fleming's Nobel Lecture and the introduction of penicillin in British hospitals, that proportion had already risen to nearly 40 per cent. Alarmingly, there were even patients who had never received penicillin who carried a resistant strain – they had missed out on the benefit of the new wonder drug. This was long before the invention of the term 'microbiome', but Barber knew that staphylococci were cosmopolitan – they are carried at any one point by a third of people on skin or in the upper respiratory tract and are easily transferred. She included a table in her article tracking a resistant strain as it slunk between different patients on a maternity unit: from one patient's breast milk to another's sticky eye.

These observations confirmed Fleming's point that antibiotic resistance was a social problem: resistance that arose in one patient could risk the life of another. But was the rapid rise in resistance due to thousands of negligent Mr Xs, or something else? The only way to reliably measure resistance levels was with bacteria in the laboratory rather than in a patient. Fleming himself proposed a method for doing so, but Barber spotted a problem with it. His method ignored the size of the inoculum of bacteria – the initial scoop of bacteria that the scientist uses for the experiment – because Fleming claimed that variation in this amount didn't affect the results. He believed that the number of bacteria didn't affect the emergence of resistance; it was simply how much antibiotic you exposed them to. That explained his view that as long you as you used enough penicillin, you could prevent resistance.

But he was wrong. As Barber showed with her own results, the size of the inoculum was crucial in determining the resistance level you measured. It mattered because a patient typically contained many millions of bacteria: if you took a smaller number of cells, your laboratory experiment wouldn't reflect reality. Fleming's method could give misleadingly low values for the dose of penicillin that would eliminate the bacteria. If you took a tiny scoop, fewer than 1,000 cells, your results would suggest the strain was not highly resistant. But a few million from the same sample – more accurately reflecting the infection inside the patient – and you would find bacteria that could withstand a concentration of antibiotics 800-fold higher. Barber's results demonstrated that resistance was far more of a problem than Fleming's methods suggested.

The truth was that penicillin itself didn't create resistant bacteria – whether they existed before treatment or arose by chance during it, penicillin merely made them visible by killing off everything else. That may sound like a purely semantic difference, but the error led to an unfortunate misunderstanding. Though low doses of antibiotics can indeed provide a pathway to resistance, the belief that bacteria were 'educated' through exposure meant that doctors began to conflate the strength of the dose with the length of an antibiotic course – as if a longer treatment was always better for reducing resistance. This motivated the well-known advice to always finish the course of antibiotics to prevent resistance, when in fact for most cases there is no good evidence that stopping treatment early encourages its emergence. (An exception is where the total eradication of a pathogen is the goal, such as in tuberculosis.) Patients trusted that using antibiotics as doctors recommended was a way of making resistance impossible, but it wasn't. Fleming's story of Mr X has had a long afterlife. Pamphlets in hospital waiting rooms still warn of 'clever' bacteria that 'learn' to resist antibiotics. Such figures of speech might seem like a harmless simplification, but the danger is that they mislead us into thinking of the problem as a war with bacteria that we might conceivably win through greater antibiotic force, rather than an evolving system that is unavoidably shaped by that force, its landscape shifting under our feet.

After the end of WWII, the spread of penicillin continued, a piece of good news in a world shattered by war. But across the world, doctors replicated Barber's observations of growing resistance. The more they used penicillin, the more resistance they found. Yet this growing problem was far from the minds of most, for whom penicillin remained miraculous. One doctor even speculated that the waters at Lourdes might owe their curative powers to penicillin. Expectations of what medicine could achieve had once again been transformed, and most people had little inkling that this brave new world might be temporary. Their confidence was hard to shake. As one Spanish bullfighter explained to a journalist, 'Thanks to Fleming, I can stand closer to the bull'.

Penicillin presents an irony: because it was so effective, it became so ubiquitous that its actual benefit was impossible for doctors to assess. The British Army's own medical reports credited penicillin as a large factor in producing a recovery rate of 93 per cent in soldiers at Allied medical units. But the antibiotic was so widely used that it was impossible to conduct the statistical comparisons that originally had been envisaged to test its efficacy. It was like running a clinical trial with only one treatment. If penicillin's true power was uncertain, what nobody could deny was how concentrated it was. In the entire D-Day operation, medical units used 32 million yards of bandages but only 39 kg of penicillin – less than the weight of a single soldier. That concentrated power of a naturally occurring antibiotic had never before been achieved. The secret had been exploiting the natural variation possible in *Penicillium* – including by inducing it with nuclear radiation – and expertise in deep-tank fungal fermentation.

The penicillin manufacturing programme is often cited as one of the first examples of 'big science', where scientific research is accelerated by huge collective efforts rather than small research groups. It took place alongside another gargantuan wartime endeavour: the Manhattan Project. There are striking similarities between penicillin and the atomic bomb. The Manhattan project also had to obtain sufficient quantities of a naturally occurring but vanishingly rare material – an isotope of uranium present at less than 1 per cent in natural uranium – in order to use its concentrated power. Even eighty years on, penicillin and the atom bomb remain powerful symbols of the promise and peril of science.

After the war ended, the Americans were anxious to retain control of both technologies. Other countries wanted their own penicillin plants, but the Americans tried to stop them. One CIA report warned that the microbiology expertise needed to make penicillin could also be used for bioweapons; the more powerful driver was simply not wanting to squander a clear technological advantage. The American state prevented some countries from obtaining penicillin, blocking exports of the medicine and refusing to issue customs forms for key manufacturing equipment made in the US. Smugglers profited from the high global demand: thanks to penicillin's concentrated form it could be easily smuggled into the Soviet Bloc through Tangiers or Switzerland, where it commanded high prices. Despite US efforts to impose restrictions, the CIA noted 'a brisk trade in antibiotics from the West to the East'. The issue caused fractures. At a conference held by the recently established World Health Organization (WHO), some countries threatened to leave the organisation if they didn't get the equipment they needed to build their own penicillin plants. A frustrated Ernst Chain, who had been the first to isolate pure penicillin, wrote a pamphlet explaining how to set up a penicillin plant. He distributed it to any interested country, because he believed everyone had the right to access penicillin.

The battle over credit for penicillin became intensely nationalistic: the British complained bitterly that the Americans had stolen the glory, and promoted Fleming as penicillin's sole discoverer. In 1944, *Time* magazine featured a magisterial Fleming gazing out in front of flasks of seething broth: the cover proclaimed that 'his penicillin will save more lives than war can spend'. But penicillin didn't belong to Fleming. His observation was brilliant, but he never even isolated the molecule – penicillin was simply what he called his impure mould juice. Penicillin didn't belong to any other individual either. Its transformation into a medicine required not simply a handful of boffins but thousands of workers. How to apportion credit for such a collective enterprise is a challenging question. But we shouldn't forget that the first entity to manufacture penicillin was the *Penicillium* mould itself. Admittedly, a mould cannot receive a Nobel Prize. But even this may not be the end of the story: recent genetic analysis suggests that the cluster of genes that makes penicillin is probably of bacterial origin. If this is true, then penicillin was a tool borrowed from the genomes of the bacteria that crowded around the hyphae of *Penicillium* in the soil. Whether you view this as a sharing or a stealing, given penicillin's contested history it seems somehow appropriate that for nearly a century, we may have been crediting the wrong kingdom of life for its evolution.

As a substance, penicillin had no legal owner. Because of the wartime circumstances under which it had been developed, no company had a monopoly. In the post-war world, penicillin's price crashed: a medicine that had been worth more than gold soon cost less than the bottle it came in. Companies that had invested heavily in penicillin were left stinging from the promise of enormous profits that had suddenly evaporated. But they had glimpsed a deeper truth: nature contained powerful medicines. Rather than relying on an obscure report of laboratory contamination with a mould from London dust, the glimmer of hope for the future was that a coordinated search could reliably recover new antibiotics as powerful as penicillin, but with some prospect of securing a monopoly. The companies would start their own search for antibiotics; they would seek the next penicillin, and they would assert their ownership over it.

The approach they would follow would be as systematic as Fleming's had been haphazard. For whereas penicillin had been a chance discovery, a new American antibiotic justifying this new approach had recently been discovered in the laboratory of a little-known microbiologist, a Ukrainian immigrant who had fallen in love with soil.

Earthly powers

Streptomycin

The soil is not a mass of dead debris . . . it is teeming with life.

Selman Waksman and Robert Starkey

For Proust, childhood was the taste of a sweet crumb of a madeleine biscuit dipped in lime tea. For Selman Waksman, it was the smell of soil. Waksman was born in 1888 in the Russian town of Nova Pryluka, now in western Ukraine. In his own words, it was 'a mere dot in the boundless steppes'. Neither agricultural nor industrial, it had nothing special about it: just a 'little godforsaken town' with harsh winters. Years later, as a graduate student living in the mild climate of California, he was surprised to find himself yearning for deep snows and heavy frosts – but more than those, it was the smell of the earth that he remembered. 'That odour of the black soil so filled my lungs,' he wrote, 'that I was never able to forget it.'

That soil was chernozem: rich and black, a soft crumbly mixture between velvety Guinness and thick chocolate cake, oozing and moist. Almost a quarter of the world's supply was located in Ukraine. The fertility of chernozem was legendary. Crops could spring from it year after year, delicate green shoots in a sea of black earth, without its nutrients ever being exhausted. A blanket of chernozem kept the Russian steppes warm: a great earthen quilt stretching across the Asian continent for thousands of miles, from the Crimea to Siberia. Under its surface the cycle of life was always churning. Chernozem is rich in humus, the top layer of soil formed by the decomposition of organic matter, which here could reach up to two metres deep. You could plunge your arm in up to the shoulder, and further still, enough to submerge your whole body. Imagine that you could breathe chernozem. You could start at the easternmost edge of Siberia, breaking the frost on the ground before slipping into the silty darkness. Once underground, you could swim westwards across most of Russia's landmass without putting your head above the surface, the human world only muffled shouts above your head, before emerging with an earthy splutter on the shores of Nova Pryluka.

If the earth has a heart, chernozem is where it beats strongest. The yearly cycles of growth and decay fold themselves back into its richness, the metabolic ebb and flow of a nutrient tide in a vast inland sea. That cycle is too slow for human eyes to see, but speed up time, and the dynamism would be revealed: ground squirrels darting like furry dolphins, bursting through shoals of minnowing worms, chased by the cold bite of winter down to the deeper soil before returning to the surface for summer's marvellous warmth. Under the young Waksman's feet, yearly migrations fretted and frothed the soil, smoothing the decay of matter into a loamy cream: the coiled trails of worms and the churning whirls of squirrels burrowing their *krotovinas*. Waksman would walk along the furrows, picking up worms between grimy finger and thumb and watching the plants force their way up. The rye grain that sprang from the chernozem would be ground and milled then transformed by a hot clay stove into a hunk of treacle-dark bread, almost the shade of the soil itself.

Given this background, it was predictable that Waksman would remain fascinated by soil. Understanding its microbial diversity would become the focus of his scientific career. What was less predictable was that this diversity contained a cure for one of the world's most deadly infectious diseases.

Waksman never intended to save the world. As a teenager, he wanted to become an agricultural scientist. He first studied at Odessa but emigrated to America in 1910, driven away from his home country by antisemitic pogroms. In America, he took his first course in general bacteriology. Looking down the microscope made him feel dizzy: a tiny sample of soil contained 'more living things, with a greater variety of forms and kinds, than there were people in a great metropolis like New York'. It was an almost identical epiphany to the one Leeuwenhoek had experienced over two centuries earlier.

By the time Waksman was studying, many bacteria had been catalogued – perhaps too many. The bacterial zoo was growing unruly. The bible on the topic of classification, published in 1900, had been *A Manual of Determinative Bacteriology*. Like a 'choose your own adventure' book, it directed the scientist to flip back and forth through flowcharted pages to end up with the right name for the organism under their microscope. It was quickly obsolete. In 1915, one American bacteriologist gloomily noted that 'to state that the classification of bacteria is in a chaotic state is to express a truism'. Anybody attempting to rationalise the sprawling mess quickly found that supposedly different species could be one and the same, while others had no classification at all. With the requisite skill and patience, a bacteriologist who looked carefully could discover not just a few species, but whole new branches of the tree of life.

Here, Waksman's visual memory was an asset. As a child in Nova Pryluka, he had been coming back from the market one day carrying his baby sister Miriam, when he was accosted by a rabbi who dragged him into the synagogue. A scribe had just completed a Sefer Torah, a handwritten copy of the Torah. With over 300,000 individual Hebrew characters, some errors are inevitable. A rabbi would examine every word. Where one looked as if it could contain an error, a longstanding tradition held that the best test of accuracy was to get a child to read a passage out loud. The child's innocence meant they could be trusted to read what was really there, and any slips in the scribe's handwriting would cause the child to read out the wrong word. For example, in the Hebrew alphabet, the fifth and eighth letters are easily mistaken for each other (ה and ח). Unfortunately for the rabbi, the eight-year-old Waksman was in a hurry to get home – and he was no innocent. Made to study the Torah at school, he had already memorised large chunks of it. Rather than deciphering the scribe's handwriting, he recognised the passage then recited it from memory. Misled, the rabbi let him go.

Like the Hebrew alphabet, the appearance of bacteria could be deceptive. The mark of a good bacterial taxonomist was to be able to classify the uncertain forms seen down the microscope, as if through a glass darkly. The skill was to see past the specks and tangles of floaters in the vitreous humour of your eye, past the imperfect appearance of a particular bacterial cell, and discern the characteristic form of the species. These methods were a craft, and Waksman had the makings of a master craftsman.

Waksman's gaze down the microscope was not his first encounter with bacteria. When he caught scarlet fever as a child in the 1890s, his anxious mother kept vigil by his bedside. Neither Waksman nor his mother would have been able to see the *Streptococcus* bacteria causing his illness. Only thirty years or so had passed between the naming of the bacteria and Waksman's infection, but even knowing the underlying cause wouldn't have helped one bit with treatment. This was the pre-antibiotic era; his survival was due more to chance than to anything his anxious mother could do. Then, when Waksman was nine, his life was changed by another species: *Corynebacterium diphtheriae*. These 'club-shaped' bacteria cause diphtheria, a disease whose name comes from the Greek for 'leather hide'. He wouldn't have seen the microscopic bacteria swarming and multiplying inside his baby sister, but he saw their effects. Miriam, aged two, fell seriously ill.

Diphtheria is a horrible way to die. The bacteria grow in the victim's throat, carpeting it with a noxious grey film – the 'leather hide' that gives the disease its name – that eventually suffocates. The human body produces antibodies, but in young children this defence is often insufficient. At the time, the only treatment that worked was serum therapy: the antitoxin manufactured by injecting horses with *Corynebacteria*, then transfusing that blood into children.

When Miriam fell ill, the nearest supply of antitoxin was in Kyiv, 200 miles away. A shipment was urgently sent for, but it arrived too late. Miriam died in the family home. Waksman was devastated. Writing decades later, his grief remained. 'I loved my little sister!' he wrote, 'I needed her so much.' Grief gnawed away at him, changing his understanding of his story, working and reworking it, until – in later life – he would state that Miriam's death had laid the foundations for his desire to conquer bacterial infection. But if that was his desire, then choosing to study soil was, on the face of it, a bizarre decision.

Waksman had selected an unpopular corner of the bacterial world. Most research into microbes concentrated on those that caused disease: the pathogens – literally 'producers of suffering' in Greek. Many of the culture methods that were so useful for pathogens, such as culturing on meat extracts, bouillon jelly or other wobbling simulacra of the body's interior, were useless for other sorts of bacteria. The pathogens were more stimulating for students, but that didn't make them the best introduction to general bacteriology. As the author of one textbook complained, you would hardly teach chemistry for beginners using dynamite – yet bacteriology taught only about the most dangerous bacteria, leaving students with warped ideas. But Waksman knew that microbes controlled all the processes of decay and regenesis: on their actions rested plant, animal and all human life.

On arrival in the States, Waksman went to work on his uncle's farm in New Jersey. He enrolled as a student at Rutgers Agricultural College nearby. His passion for soil microbiology propelled him through college. He had less interest in his other subjects and cut corners where he could. Forced to write a compulsory English essay each quarter, he chose to write on the same topic – 'The insulted and repressed types in Russian literature' – sixteen times in a row. Any literary skill he possessed was turned to a practical bent: aged twenty-two, his article 'How I Raised a Flock of Chickens' for the *Rural New Yorker* won him $10. He graduated in 1915, moved to California to get his PhD in 1918, and returned to Rutgers as a soil scientist.

He established his reputation as a world-leader in the small field of soil bacteriology by becoming a microbial census-taker, cataloguing the diversity of soil. Like other soil bacteriologists, he was sometimes almost proud of the irrelevance of his subject to infectious disease. He revelled in the different methods needed to study soil bacteria compared to pathogens. The idea that there could be any sort of connection between the two was preposterous. It was at a conference on soil in Rome in 1924 that he met a young scientist called René Dubos who would change his mind.

René Dubos was a French scientist, born in 1901 in a small village just north of Paris. He had been a fit child, but at the age of eight, after winning a bicycle race, he came down with a throat infection – likely caused by streptococci – which festered into severe rheumatic fever, an inflammatory disease with acute joint pain that lingered like a chronic aftertaste. This was twenty-five years before Prontosil. Dubos didn't die, but he suffered the next-worst thing for an eight-year-old: eighteen months confined to his bed. The infection left him with a weakened heart for the rest of his life. Such an early encounter with bacteria might have been expected to inspire him to pursue microbiology. Not so: the detrimental effects of his illness on his schooling meant that he only passed the exams to attend a single institution, the Institut National Agronomique, where students learned agronomy – the science of soil management and crops. So, through microbial contingency, Dubos found himself an unlikely student of scientific farming.

For colonial European nations, cultivation was civilisation. This mode of thinking suggested that indigenous people without recognisably European agriculture didn't truly own the land they inhabited, conveniently leaving it ripe for the taking by the first Europeans who planted crops. Enhancing the fertility of soil was one of agronomy's key obsessions. Karl Marx had argued that the basis of the capitalist mode of production was 'the expropriation of the great mass of the people from the soil'. Agronomists aiming to increase soil fertility were, in Marx's view, disturbing this historic relationship and 'robbing the soil'. That made agronomy sound far more exciting than Dubos found it. He was attracted by the opportunity to join the ranks of colonial administrators and leave France for French Indochina. But it wasn't to be. The ghost of his early infection was still haunting his body, and during his officer training at a French Army school, his poor health led to him being disqualified. In February 1923, having graduated, he instead found himself trapped in Europe, with a job at the International Institute of Agriculture in Rome.

Rome was a troubled city. The National Fascist Party had recently swept to power in an effective coup, marching on the city in October 1922 to force the installation of Benito Mussolini as prime minister. Under Mussolini, the fascist regime worked to increase the expanse of the Italian colonial empire. The Institute of Agriculture, founded in 1905, was an international effort to globalise farming methods. Reports came in from all over the world, and Dubos would distil them into pithy, accessible abstracts to be redistributed. Dubos still had no discernible interest in bacteria: he wrote about sturdy topics like ploughs, harvesters and land management. Then one day, scouring through reams of articles as part of his routine work, he came across a piece by Russian scientist Sergei Winogradsky that called for a new field of microbiology – a dynamic microbiology of soil.

In contrast to the way agronomy treated soil as a resource to be exploited, Winogradsky wrote about it with a reverence that bordered on the religious, a heaven thronged with an angelic chorus of bacteria. 'Soil is a living world,' he wrote, 'saturated with a teeming mass of microscopic beings, having a variety defying all imagination'. In the late nineteenth century, Winogradsky, a Russian aristocrat who had trained as a concert pianist, had raised the mysterious underworld of soil above ground. He filled columns of glass with earth and watched as the community separated into different coloured layers, like stripes of Neapolitan ice cream. Aged just forty-nine, Winogradsky had then retired from research because of ill health. As a wealthy landowner, he spent the next seventeen years managing his vast agricultural estate and playing the piano. But the Russian Revolution of 1917 made clear this feudal lifestyle was dangerously time-limited. In 1921, he fled from Odessa on a French warship to Marseilles, then on to Switzerland, Belgrade and finally the Pasteur Institute in Paris, which granted him a new laboratory of his own on a leafy estate outside the city. He was based there when Dubos first read his work.

For Dubos, schooled in the arid and practical soil of colonial agronomy, Winogradsky's arguments were bewitching. The earth was alive: it had, according to Winogradsky, the 'characteristic functions of a living organism'. Just a few weeks after Dubos had first read Winogradsky's article, there was to be a conference on soil held in Rome. It would have been the perfect chance to meet his new hero. But amid the throng of the conference, the closest Dubos got was a sighting of the tall and imposing Russian from twenty feet away.

Fortunately, Dubos was able to meet other soil scientists, including Waksman, who encouraged him to pursue his newfound passion for soil bacteria and come to America to work at Rutgers. That summer, an energised Dubos crammed a course in bacteriology and then bought a one-way ticket on the *Rochambeau* from Le Havre to New York. Though Waksman barely knew Dubos, he had offered him the chance to come and work at Rutgers on soil science. It's possible that he'd overheard what another American colleague had reportedly said in Rome: 'I know a winner when I see one.'

In his new incarnation as a soil microbiologist, Dubos became fascinated by the self-purifying nature of soil. It was an almost trivial observation, but soil was capable of absorbing and decomposing any living matter – no matter how tough – if given enough time. The mightiest trees could be dissolved by tiny microbes. At Rutgers, Dubos started by working on the soil's decomposition of cellulose, the material that strengthens the cell walls of plants. One day, he was visiting the Rockefeller Institute in New York when at lunch he sat next to a medical researcher called Oswald Avery, who talked about the difficulty of treating bacterial infections. Dubos had a flash of insight that connected their two worlds.

Many bacteria that caused infections were difficult to destroy because they possessed the equivalent of a cellular shield. One example would be *Streptococcus*, a genus of bacteria including the species that had given Waksman scarlet fever in childhood. Certain *Streptococcus* species were known to have virulent forms possessing a tough outer layer made of sugar molecules called polysaccharides. That 'capsule' was evolved to withstand attack by the immune system. Yet there was no such thing as a biological material that couldn't be degraded: *Streptococcus* had existed for millions of years, but the ground was not littered with its capsules. Therefore, like all living material, the capsule must be recycled in the soil. That meant, said Dubos, that somewhere there was a microbe that could destroy it – and that knowledge could be worked up into a medicine.

The assertion was as brazen as it was bizarre. Dubos, a soil microbiologist, was claiming that he could succeed where the world's best doctors had failed. Avery took the idea seriously enough to offer him a job at the Rockefeller. Dubos would attempt to learn from soil bacteria how to destroy *Streptococcus*.

Dubos started with soil from a cranberry bog. Cranberries prefer acidic conditions: their ideal pH is more than 4.2 but less than 5.5 – not as tart as a lemon, but somewhere between tomato juice and black coffee. He adapted the methods he knew from Winogradsky: he created a microcosm of the cranberry bog in his laboratory. Then he began a process of enrichment, feeding his miniature bog a purified extract he had produced of the *Streptococcus* capsule. As long as he added no other material, over time the only bacteria that survived would be those that had been able to consume the capsule. He watched with amazement as his idea worked. He'd found proof of his general hypothesis: whatever you wanted decomposed, something in soil could do it. In time, he even isolated the enzyme responsible for the digestion of the capsule, but the enzyme proved difficult to purify and use in humans. It was a technical proof that he was right, but not the transformative discovery he had hoped for.

Indeed, after the discovery of Prontosil in 1935 and the resulting hundreds of sulfa drugs manufactured by chemists in their laboratories, soil must have seemed a disappointment. It would have been easy for Dubos to abandon his slow work digging around in cranberry bogs, an amateurish pursuit in comparison to the might of modern industrial chemistry. But he persisted. In 1939, he made a new and astonishing discovery. He was studying *Bacillus brevis*, a harmless Gram-positive bacterium that was common in soil, feeding it not the capsule but whole bacteria in suspension. He found that it could kill other Gram-positive bacteria, somehow dissolving them with a molecular poison. Dubos would cook up gallons of *Bacillus brevis* then reduce the mixture, like a chef preparing stock, creating a sticky brown mess – a colleague recalled it looking like 'uncouth earwax'. The muddy earwax contained the poison, which Dubos called gramicidin. He estimated that in mice with bacterial infections, gramicidin was one hundred thousand times more powerful than the sulfa drugs.

At a conference of microbiologists in New York in September that year, Dubos presented his discovery. Scientists rarely use props, but, like a child doing show-and-tell, Dubos revealed a small vial of gramicidin and explained to his audience that it contained enough medicine to treat 5 trillion mice. It was a preposterous statement. Dubos's vial defied common sense – and yet, there it was.

Gramicidin proved not to be the miracle that many had hoped for. It was too toxic to human cells to be safely used inside the body, though Dubos never lost his admiration for its powers (he even took to sprinkling it on his own garden). But it was still a counter-intuitive spark that started a fire. If the idea of medicines from microbes had been smouldering before in the minds of a few others in his audience, now the blaze caught. The oxygen of publicity helped. One newspaper article described Dubos's work as 'like killing a tiger, not merely by shooting him with a bullet . . . but by pouring over him a solution that reduces him to a puddle of decomposed flesh'. The audience in New York that day included Howard Florey, who returned to Oxford and stepped up the Dunn School's research into natural products such as penicillin. Also present was Dubos's previous mentor in soil microbiology, Selman Waksman. The month after Dubos's demonstration, Waksman published an article where he noted that disease-causing bacteria must have been passing from people into the soil for thousands of years, but were completely missing when scientists looked for them: 'What has become of all the bacteria causing typhoid, dysentery, cholera, diphtheria, pneumonia, bubonic plague, tuberculosis, leprosy, and numerous others?' Perhaps the resident soil microbes were capable of destroying the killers, as Dubos had suggested.

Waksman recognised that for too long there had been a mutual incomprehension between soil and medical bacteriologists. Knowledge of these two different worlds had developed more on parallel tracks than in symbiosis. Now, Waksman followed Dubos and crossed those disciplinary wires, outlining protocols for culturing soil and human bacteria together. These methods were ways of winnowing the great diversity of soil down to the species that could defeat a pathogen. Waksman trusted that if he asked the question enough times of soil, the answer would eventually be written obscurely in the visual language of the agar plate.

Following Dubos, Waksman was putting faith in the microbes and their ability to make molecules that could be turned into medicines. The search was necessarily vague: not a treasure map where 'X' marked the spot, but a whole uncharted continent. The only thing that could be said in generality was that the soil contained molecules made by life that acted against other life. A French scientist in the late nineteenth century had coined the term *antibiosis* for the general phenomenon of 'life against life' – from lion against zebra, to tortoise against lettuce. From this, Waksman would eventually formulate a name for the desired molecules: they were antibiotics.

At the start of 1939, the pharmaceutical company Merck employed Waksman as a consultant, and soon he was earning over $250 a month. Waksman predicted that penicillin and gramicidin were not unusual: he would soon discover other molecules that fought disease. Merck proposed a mutually beneficial arrangement. Waksman would share new substances discovered in his laboratory at Rutgers, and in return Merck would share expertise and run tests and clinical trials. Under the agreement, signed in November 1940, Merck would get exclusive rights to manufacture and sell any antibiotic that Waksman's group discovered, paying Rutgers a small royalty of 2.5 per cent on net sales.

By 1943, there had been no successes, but two near misses. Both new antibiotics looked promising, but one killed all the laboratory rats, and the other worked in mice but caused kidney failure in humans. And yet that summer, a stone's throw from where Waksman had raised the flock of chickens that had won him a $10 prize, his prediction would come true. But it would not be his discovery.

Like Waksman, Albert Schatz was of Russian-Jewish heritage. He had grown up on a farm in Connecticut – one summer he slept on a cot in a chicken coop – and had always planned to return there. But after going to Rutgers as an eighteen-year-old in 1938 to learn agriculture, he had been seized by the soil in a similar way to Dubos. In May 1942, he graduated top of his class with a major in soil science. His favourite professor had no funding for him, so he ended up with his second choice: a PhD in microbiology with Waksman searching for antibiotics. After only a few months, he was swept down to Florida by army service in WWII for almost a year, working in an army bacteriology laboratory with pathogens. But he chose to return. With a meagre PhD stipend, Schatz saved money by sleeping rent-free in a room of the university greenhouse in return for watering the plants and eating produce grown in agricultural experiments. He was almost always hungry, and he had little time for anything outside his research. Rather than testing the unpromising candidates Waksman had already found, he wanted to search for a new antibiotic of his own.

One killer loomed large: *Mycobacterium tuberculosis*, the pathogen that caused tuberculosis. Tuberculosis, also known as consumption, was a slow-burning disease. Contracted in childhood, *M. tuberculosis* would lie low for years. It was a formidable foe, supremely indifferent to sulfa drugs and penicillin, protected by a waxy outer layer that guarded it like a capsule. Its ancestors had come from the soil, but had slowly adapted to live inside human lungs, floating between friends and family in coughed-up droplets. Tuberculosis could take several forms, but its most well-known was pulmonary tuberculosis, which affected the lungs. The bacteria formed clumped nodules inside the victim's chest, sequestering themselves away from attack by the immune system. Sufferers could appear healthy until they began to cough up blood and waste away. Though infections had reduced dramatically thanks to improved housing and sanitation, tuberculosis still killed more than 50,000 Americans each year. Doctors had been trying for decades to find a reliable treatment without success.

The field was littered with failure. Robert Koch, the great German microbiologist, first claimed to have discovered a cure in 1890. His purported cure was prepared from cultures of *M. tuberculosis* itself, mixed with sugar alcohol to produce a substance called tuberculin that was injected into the skin. Initial excitement and jubilation turned to despondency as it became clear that tuberculin did not cure patients, and Koch's reputation was permanently damaged by the embarrassment. Other researchers had tried to develop vaccination, with little success: the best vaccine, formulated by two French researchers in 1921, was less than 20 per cent effective.

Waksman had shown as early as 1931 that *M. tuberculosis* could survive for months in sterilised soil. But in unsterilised earth it disappeared, suggesting that other soil bacteria were attacking it. He never identified those bacteria, but when he tasked Schatz with investigating new antibiotics, Schatz tested soil bacteria against a range of pathogens including a harmless strain of *M. tuberculosis*.

Schatz became excited about the killing abilities of two strains of *Streptomyces griseus*, a bacterium that he had isolated first from farmyard soil spread thickly with manure, and then again in a swab taken from a chicken's throat. In October 1943, Schatz confirmed that *S. griseus* could kill many of the pathogens on his hit list. *M. tuberculosis* was one of them. Another scientist in Waksman's laboratory, Elizabeth Bugie, managed to isolate and purify the antibiotic responsible, and they called it streptomycin.

In February 1944, Waksman sent an advance copy of the paper to doctors at the prestigious Mayo Clinic, who were amazed that streptomycin could kill *M. tuberculosis*. They immediately encouraged Waksman to test it against a deadly clinical strain. Accustomed to working with soil bacteria that posed no risk to humans, Waksman blanched at the idea of dangerous strains of *M. tuberculosis* in his regular laboratory. This was not an irrational fear – many researchers did indeed contract tuberculosis from the very samples they were working on. Schatz, however, was undeterred. As a child, he had watched the funeral procession of a man who had died of tuberculosis. When the black hearse passed by, pulled by two black horses, the man's widow had run after the carriage and clawed in desperation at the doors. By March 1944, Schatz was working with such deadly strains himself, protecting himself by rinsing out his mouth with antiseptic wash. Sometimes, he would sleep overnight in the basement where he conducted experiments; he even invited his girlfriend there for occasional dates. Waksman, in contrast, remained 'deathly afraid' of the disease, according to Schatz, and never once descended to the laboratory once the experiments began.

Results from streptomycin were promising, but the Mayo Clinic doctors wanted to test it in real patients. That meant Schatz needed to make more of it. The basement started to resemble an underground brewery, two or three glass stills constantly bubbling away with soil bacteria mixed with solvents. It reminded Schatz of the illegal alcohol he'd made as a young boy during Prohibition. He slept next to his streptomycin stills so he could keep replenishing them, the building's night watchman prodding him awake if the liquid level fell below a red pencil mark.

Had Waksman ever poked his head around the door of Schatz's basement, the smell might have been familiar. As well as streptomycin, *Streptomyces griseus* and other related soil bacteria make a chemical called geosmin which is the main contributor to soil's earthy smell. The aroma came from the same bacteria that were about to change medicine – and make one of the two scientists a rich man.

As streptomycin's promise became more apparent, the question was whether its production could be scaled up in the same way as penicillin. But although the US government's War Production Board considered the possibility, the situation was different: no government funds had yet been spent on streptomycin and it owned no rights to it. That was despite the discovery having been made in Rutgers, a public university – because Waksman had signed away all future rights to Merck. Waksman was having second thoughts about that agreement, which was looking increasingly unwise. How would it appear if it emerged that he had given away a life-saving medicine to a company which now had 'absolute control' over its use? In June 1944, he asked Merck to abandon their agreement, citing the potential value of antibiotics to the armed forces and the wartime emergency.

Merck's CEO, George Wilhelm Merck, was an imposing presence: six foot five and 250 pounds. Merck was still a family affair, and his father had been the CEO before him. He had overseen the growth of Merck's research arm, employing academics directly and also partnering with them, as in the research agreement with Waksman. Now Waksman was asking Merck to voluntarily give up its exclusive right to streptomycin. Remarkably, George Merck agreed. Rutgers and Merck struck a new deal. Instead of being assigned to Merck, the patents from streptomycin would be assigned to a dedicated research foundation at Rutgers. From now on, the Rutgers Foundation would handle all patents from Rutgers employees. Merck would return 2.5 per cent of net sales to Rutgers, with the difference being that its licence would no longer be exclusive; Rutgers could choose to license streptomycin to other companies as well.

Although Merck appeared to have voluntarily abandoned its profits for the greater good of US war efforts, the truth was more complicated. Merck did not walk away with nothing. Waksman had been foolish in signing away the rights to all antibiotics discovered at Rutgers; now Rutgers would have to pay for that folly. Merck demanded repayment for the research it had carried out already, which after haggling was agreed to be $500,000. George Merck was not a fool, and he was also head of the government's War Research Service. Though Merck would have been legally entitled to enforce its monopoly and refuse to license streptomycin to other companies, such a move would be unpatriotic. Indeed, the US government would almost certainly have felt the need to step in and strip it of its rights, potentially turning streptomycin into another penicillin. By renegotiating their deal with Rutgers, Merck at least got paid back for their investment – and ensured that Merck's lawyers would be involved in drafting the patent application. That was crucial because, as Waksman had experienced first-hand, obtaining a patent on an antibiotic made by soil bacteria was not trivial.

The reason was that patents were reserved for inventions, not discoveries. A synthetic chemical could be patented. But a long-standing rule in US intellectual property law held that any 'product of nature' – a substance that occurred naturally – was not eligible for a patent. This argument had been used by the US Patent Office to reject the patents Merck and Waksman had filed for the two previous failed antibiotics discovered at Rutgers: a product of soil bacteria was surely a product of nature, not an invention. However, Waksman's responses to the patent examiner had pushed back against that claim, with help from Merck's lawyers. He argued that the antibiotics described were not found in soil and needed the methods described in the patent to exist at detectable amounts. The Patent Office had eventually approved the patents, although Merck had then been unable to profit from them because those antibiotics had proved unsafe.

The legal precedent that antibiotics were not necessarily 'products of nature' was tantalising. Prontosil and penicillin had both lacked patent protection in crucial ways. Prontosil had been an own goal by Bayer: by unwittingly patenting its active ingredient (sulfanilamide) in 1908, they had robbed themselves of any patent protections when its much more valuable application as a medicine was discovered. The explosion of sulfa drugs in the aftermath of Prontosil was helped by that lack of a patent: hundreds of new derivative drugs with similar effects could be made and patented, and no company had exclusive rights on the central molecule that made all this innovation possible. With penicillin, a combination of chaos and wartime collaboration had again produced a situation where there was no patent on the molecule itself. Though patents existed on particular manufacturing processes, no company had sole ownership of penicillin.

Merck had just given up their exclusive rights to streptomycin, but those rights were hypothetical: despite the new precedent, it remained an open question whether Rutgers would even be granted the patent. Waksman would have certainly been aware of the need for Merck's expensive corporate lawyers on hand to argue his case.

In September 1944, approximately a year after streptomycin had first been discovered, a secret memo arrived on Waksman's desk. As Waksman opened it, he must have anxiously scanned the details. A doctor had written of a case of a sick two-week-old baby who was crawling with bacteria. An infection had started in the child's urinary tract and moved into the blood, before swarming into the membranes that surrounded the brain to cause meningitis. The child had received the two latest drugs for bacterial infections: a sulfa drug called sulfadiazine and penicillin. They had some effect, but the infection continued to progress. The child was at the point of death – but then the doctors tried streptomycin, delivering injections of the experimental drug every three hours. A few days of treatment coincided with the ebbing away of the infection, and the baby survived.

As Waksman put down the letter, his mind must have raced. With this report and other experiments in humans, streptomycin's powers were becoming clearer. But rather than seeing it as a culmination of collaboration, emerging from ideas that had flowed freely around the scientific community and from work performed by multiple researchers, he saw it as a personal triumph. Though he had once refused to enter Schatz's laboratory, Waksman now strove to present himself as streptomycin's sole discoverer. As he would one day write, he saw the researchers who toiled in his laboratory as 'the fingers of my hand'. It was a telling metaphor: Waksman saw Schatz as an appendage of himself. In another analogy that

Waksman was fond of repeating, his laboratory could be compared to an orchestra: Waksman was the conductor, 'leading and assigning the task to each member, none of which would have produced any symphony otherwise'.

There was, of course, a problem. The original streptomycin paper had three authors: Schatz, who had done most of the work; Elizabeth Bugie, who had first purified the antibiotic; and Waksman. As a woman in the male-dominated world of science, Bugie was easy to exclude: after all, she would surely be getting married and leaving science soon anyway. Under pressure, Bugie signed an affidavit stating that Schatz and Waksman were the sole discoverers – a decision she later regretted.

But Schatz was more complicated, and his claim to streptomycin was stronger than Waksman's. It was Schatz who had isolated *Streptomyces griseus*, Schatz who had noticed its activity, and Schatz who had performed the first crucial experiments. The month before the secret memo landed on his desk, Waksman and Schatz had signed documents in the presence of Merck's lawyers that listed them as equal co-inventors, and the eventual patent that was filed described them as such. But already by the following month, November 1944, Waksman had published a paper that claimed streptomycin had been discovered according to a strict six-step protocol of his own devising. In fact, as Waksman must have known from reading Schatz's PhD thesis, Schatz hadn't followed Waksman's protocol. The statement in his paper was a lie. Had Schatz followed Waksman's orders, streptomycin would not have been discovered.

Waksman manoeuvred further. He had persuaded an unwilling Schatz to sign the patent documents on the grounds that it was necessary to prevent another company swooping in on streptomycin. In 1946, still waiting for a decision from the Patent Office, Waksman explained to Schatz that they would now formally abandon their rights as inventors, signing documents that legally assigned the patent to the Rutgers Foundation. In return for signing, Schatz would receive a cheque for the nominal amount of $1. Waksman presented this decision as the pair giving up their claim to streptomycin for the common good. He didn't tell Schatz that he had already signed a separate agreement with the Rutgers Foundation, just three days before. Of the streptomycin royalties paid to the Rutgers Foundation, 20 per cent were to go to Waksman personally – and none to Schatz.

Schatz knew nothing of this secret arrangement. He was confused by the legal language of patents and rights, and simply did as he was asked by his trusted supervisor, even though he felt uneasy about it. After his PhD was awarded in 1945, Schatz struggled financially. In an apparent act of spontaneous generosity, Waksman sent him a cheque for $500 – followed by another one, which Schatz politely rejected, saying he was now financially stable. Yet Waksman's cheques were not generosity, but guilt.

Waksman presented himself as a saviour. He claimed to have sent out streptomycin-producing bacteria to over one hundred prospective manufacturers, signing what he called 'gentleman's agreements' where they would return 2.5 per cent of their profits to the Rutgers Foundation. By 1947, the Rutgers Foundation had licensed eleven companies to manufacture streptomycin. That meant that by the following year, more than 99 per cent of American antibiotic production was of antibiotics without exclusive patents, helping to drive prices down. Behind the scenes, Waksman himself was directly profiting from those sales. By the end of September 1948, he had received over $187,000 in streptomycin royalties, around twenty times his annual salary as a professor.

When they had signed the patent application as co-inventors back in 1945, Schatz claimed Waksman had told him that they were 'partners in streptomycin' but 'neither of us would profit financially from the discovery'. Whether Waksman had honestly believed that at the time or not, when Schatz discovered that Waksman was profiting financially, he was furious. In 1950, he took the remarkable step of suing Waksman and the Rutgers Foundation. Waksman tried to evade deposition, claiming he was too busy to spare the time, but the judge issued a subpoena. Under oath he proved an unreliable witness, changing his story and resorting to his beloved orchestra analogy to try and minimise Schatz's role. It soon became clear that Schatz's case was strong. In December 1950, lawyers for the two sides reached a settlement. Schatz would receive 3 per cent of all streptomycin royalties – and not just Schatz. The rest of Waksman's orchestra was to be paid too: twenty-four other past and present members of Waksman's laboratory received smaller percentages, even those who had nothing to do with streptomycin.

Waksman had tried to tell a story of his own genius and sacrifice: a scientific visionary who had seen what others could not, who had mobilised an army to follow his orders, and abandoned the profit motive in pursuit of a world where tuberculosis could finally be cured. Waksman was brilliant, but in every detail the reality was more squalid. He had not been the first person to see the possibility of medicine from soil. He had not given Schatz instructions that allowed him to discover streptomycin. And he had profited personally from its sales. More tragically, though streptomycin did initially seem like a wonder drug for tuberculosis, its limitations soon became clear. George Orwell, seriously ill with tuberculosis in 1948, was told by his doctors that the new American antibiotic was worth trying. At the time it wasn't available in Britain, so Orwell had to pull political strings to get some imported. Unfortunately, the drug gave Orwell horrible side-effects: he noted grimly, 'I suppose with all these drugs it's rather a case of sinking the ship to get rid of the rats.' It did him no good, and he died in January 1950.

Orwell was not alone. Doctors soon realised that streptomycin wasn't the deliverance from tuberculosis they had hoped for. As Mary Barber had shown for penicillin, its initial strength didn't last. Resistance mutations often lurked within the cloud of variation in the millions of *M. tuberculosis* cells that made up a typical infection. If resistant cells survived, they would in time regain a foothold in the body. Indeed, many patients surged from illness to health, gasping with amazement at their miraculous reprieve, only to soon plunge back into the abyss.

Schatz too sank into obscurity, despite the court ruling in his favour. In 1952, the Nobel Prize was awarded to Waksman alone. The narrative that Waksman had so brilliantly spun had convinced the world. During his visit to Sweden to accept the prize, Waksman visited the official opening of a streptomycin-producing factory in the company of the Swedish crown prince. He asked the prince to lean in and smell one of the bacterial cultures in front of them. It had the unmistakable odour of freshly ploughed soil.

Assault from all quarters
Isoniazid

My experience in the sanatorium is the best teaching I got.

Unknown Navajo health visitor

At the end of March 1951, Philip was a tall and energetic twenty-four-year-old. He enjoyed exercise. His favourite sport was squash, and for holidays he liked to go walking in the hills. It was on a walking holiday that he had first met Liz, the young woman he had just married. The couple were anxious to begin their lives together. Just days after their wedding, they had moved from London to the north-east of England for Philip's new job, so that he could start work as an engineer at a port near to the smoggy town of Middlesbrough, founded in the 1830s after the discovery of iron close to the coalfields of County Durham. But within two weeks of their wedding day, Philip began to feel unwell. Suddenly he was horrified to find himself coughing up mouthfuls of bright red blood, frothy from its journey up from his lungs, the telltale sign of a hidden sickness that had been slowly fermenting within his body for years. He had pulmonary tuberculosis.

Philip had only ever heard tuberculosis whispered about. The disease was a dread diagnosis, something at once mortal and mortifying. Franz Kafka, who died of tuberculosis at the age of forty, wrote that in discussing it people dropped into a 'shy, evasive, glassy-eyed manner of speech'. After his diagnosis, Philip's previous life receded from him. He was rushed outside Middlesbrough to purer air, to a tuberculosis sanatorium that had been established in 1932, built in the grounds of a country house called Grey Towers.

The sanatorium was a place for long-term treatment. After the Industrial Revolution, rapid urbanisation had created the perfect conditions for tuberculosis to spread, as people in search of a better life found themselves packed into cramped and airless rooms. In the early twentieth century, sanatoria had sprung up outside cities. At the start of the 1950s, despite declines in tuberculosis, many of them remained in use, skulking at the outskirts of life; in the green fringes of countryside around a city they could be found blooming like tuberculous nodules. Sanatoria had been placed there for a reason. Most cities were still powered by coal, thousands of chimneys and smokestacks belching out soot that settled over the inhabitants like gritty snow, exacerbating the condition of those with tuberculosis. Architects competed to design modernist masterpieces that empha-sised cleanliness and pure air: sleek constructions of glass, steel and concrete, concealing their true function as staging posts for the dying.

From his bed, Philip had a beautiful view of the hills to the south, unobstructed by other houses. But the secluded location wasn't just for the benefit of the patients; policed by nurses, their contagion could be kept away from the rest of society. Many who passed through the doors knew they would do so only once. The corridors were filled with the ghastly coughing of tuberculosis suf-ferers, described by Thomas Mann in his novel *The Magic Mountain* as 'a coughing that had no conviction and gave no relief . . . a feeble, dreadful welling up of the juices of organic dissolution'.

The sanatorium was purgatory. Outside of the limited visiting hours, there was almost nothing for a patient like Philip to do but look out the window or read. Propped up on pillows, he worked his way through the novels of Charles Dickens. Outside the sanator-ium, post-war Britain was slowly moving towards the future; inside, time slowed to a standstill as people struggled against a Victorian disease. Philip read in *Nicholas Nickleby* what Dickens had written about tuberculosis in the 1830s:

a dread disease which so prepares its victim, as it were, for death . . . in which the struggle between soul and body is so gradual, quiet, and solemn, and the result so sure, that day by day, and grain by grain, the mortal part wastes and withers away.

But Philip was far luckier than a Dickensian character. Had he contracted the disease just a few years earlier, there would have been nothing to treat him with. Now there was not only streptomycin, but another antibiotic that could be administered alongside it. When treated with two different antibiotics, bacteria would need separate and independent mutations that conferred the different resistances to survive; the probability of this was significantly lower.

The newer antibiotic had come from an inspired piece of reasoning. The biochemist Frederick Bernheim had no training in bacteriology, but had decided at the outbreak of WWII to work on tuberculosis, as Albert Schatz had done in Waksman's laboratory a few years later. In 1940, Bernheim had discovered that a molecule called salicylic acid stimulated *M. tuberculosis*'s metabolism. He reasoned that by supplying a similar but useless molecule, he could trick the bacteria to engage with that instead, thereby inhibiting their metabolism and starving them to death. A Swedish scientist, Jörgen Lehmann, found a molecule called para-aminosalicylic acid (PAS) that was close enough to salicylic acid to achieve this effect. In March 1944, the first patient was treated with PAS in a Gothenburg sanatorium. At first, Lehmann and his colleagues gave PAS only to hopeless cases, but as they became more confident, they extended the treatment to others. Soon, doctors around the world started using PAS. By December 1949, the British Medical Research Council had published interim results from a controlled clinical trial proving that PAS slowed the progression of disease: PAS was less effective than streptomycin on its own, but when the antibiotics were used together, the rate of resistance in patients was much lower than for streptomycin alone. Humanity was learning how to mitigate the development of resistance.

In Britain, the National Health Service had been founded in 1948, so Philip's treatment was free. He was given a daily injection of streptomycin and took pills of PAS. But these did not kill the bacteria on their own. The intended effect was only to stop the bacteria spreading in his lungs, buying him time while he was lined up for an operation called a thoracoplasty. The seventeenth-century Italian physician Giorgio Baglivi is credited with the first observation in the history of this gruesome approach, noting that a man with tuberculosis who suffered a sword wound to his chest had unexpectedly improved. Over the nineteenth century, doctors decided that a collapsed lung filled with tuberculosis was paradoxically more free to heal: the collapse was believed to slow the spread of tuberculosis, restricting both the movement of bacteria through the lung and their access to oxygen. In Philip's case, a pioneering Norwegian surgeon had invented a new method shortly before WWII, and that summer he was touring the sanatoria of Europe like a concert pianist, including Grey Towers, where he removed several of Philip's ribs under local anaesthetic. It took about forty minutes.

After a few months of recovery, in December 1951 Philip was up and out of his bed, looking for something to do. Being a trustworthy-looking chap, he was put in charge of the ward's medicine timesheets, where he recorded antibiotic doses for all the other patients. In February 1952, he was discharged after ten months in purgatory and could finally start his life with Liz. In 2025, he celebrated his ninety-ninth birthday.

From my childhood, I was dimly aware that once, long ago, my grandfather Philip had been ill enough to be in hospital for a year. The fact was occasionally mentioned to explain his prodigious knowledge of great literature. (He remains the only person I know to have read *War and Peace* twice.) But I had never broached the topic with him until I began researching this book. After I told him I wanted to discuss it, in typical fashion he made detailed notes before our first conversation, his neat engineer's handwriting jotting down the points he wished to discuss as an aide-memoire. It was only in that first conversation that I realised just how ill he had been.

As I researched the history of tuberculosis, I came to appreciate that my grandfather had contracted the disease on the right side of a historic transition: a world with a new health service, new surgeries and new antibiotics. But while using streptomycin and PAS together was a significant advance, and it worked in his case, it was not sufficient for many patients. One problem was that streptomycin was only effective against the *M. tuberculosis* cells when they were actively growing; dormant cells could reawaken to persist unaffected. Unless patients were treated early, they would live the rest of their lives shaped by the disease. That was about to change. At the same time that Philip was walking out of the sanatorium in February 1952, news was filtering out from America of an amazing new medicine being used at a sanatorium in New York.

When it was dedicated in 1913, the Sea View Hospital for tuberculosis was said to be the largest and finest ever built. Sited on Staten Island, far to the south of New York's smog, across the tidal strait where the Atlantic breeze could roll in through the windows, Sea View resembled a quaint village from an older and gentler world. Its four-storey pavilions were designed in the Spanish Mission style, fanned with pleasing symmetry around a central dining hall. Patients lay in curved sunrooms, soaking up light and breathing in the pure air as they gazed out on neat gardens. The beds were quickly filled – and still the need for more grew, as tuberculosis raged throughout the city. Through the 1920s and 1930s, the Sea View complex had sprawled out – more dormitories, open-air cottages and even a six-storey building for children only – until the site covered over 300 acres of woodland. Even at the beginning of the 1950s, a quarter of New York's hospital beds were occupied by tuberculosis sufferers. New York's Commissioner for Hospitals, Marcus Kogel, followed the progress of new treatments for tuberculosis with interest. In particular, a trial at Sea View seemed to be delivering results beyond what anybody could have believed. On 20 February 1952, Kogel held a press conference.

In front of the assembled journalists, Kogel outlined how a new antibiotic called Rimifon, manufactured at the Swiss pharmaceutical firm Roche, had been trialled at Sea View. The results with Rimifon were astounding. Indeed, they were scarcely believable. The antibiotic was curing desperate cases where doctors had previously had no hope. Wards that had been sombre antechambers for death were now the sites of pyjama parties. The patients' appetites had grown to epic proportions: some were having three bowls of cereal and five eggs for breakfast. The resulting weight gains were so dramatic that doctors realised that the amount of drug they were giving 'per weight' needed to be regularly recalculated, in order to keep a patient on the same effective dose. Journalists went out to Staten Island and onto the wards to verify the news for themselves, interviewing patients who were jubilant at the improvements in their health: a photographer from *Life* magazine took pictures of them dancing in the corridors. One patient had told doctors on admission, 'I don't believe in anything anymore. The quicker it's all over, the better.' Now he was cured. The mood in the sanatorium finally matched the sunlight that streamed through the windows.

One might have thought that the Roche scientists would share the patients' euphoria – it would be difficult to imagine a more perfect way to launch Rimifon. But Kogel had not consulted them about his press conference beforehand, and they were horrified. His announcement started a chain reaction of phone calls throughout the night, as journalists machine-gunned Roche with questions. The morning after the press conference, the American branch of Roche sent a telegram back to the firm's headquarters in Basel, Switzerland: 'Due to Kogel indiscretion US press carrying articles on Rimifon forcing us to take relevant measures'.

This was not how Roche had wanted the story to unfold.

Kogel's press conference had been preceded by a year of intense activity at the company. In December 1950, two scientists called Robert Schnitzer and Emanuel Grunberg had identified Rimifon as an outstanding drug against *M. tuberculosis*: twenty times more effective than streptomycin, it was active even at a dilution of one part in 60 million. It worked on pulmonary tuberculosis and on 'miliary' tuberculosis too, a form of the disease where *M. tuberculosis* cells enter the bloodstream and infect other organs, causing clusters of small lesions that resemble millet seeds. When mice with tuberculosis were given streptomycin and then treatment was stopped, despite appearing cured within three weeks, their organs would be visibly crowded with miliary tuberculosis. With Rimifon, the millet seeds never appeared. Roche scientists tested Rimifon in rats, dogs and monkeys with similar success.

By May 1951, the Roche scientists were satisfied that they knew Rimifon's toxicity and could recommend a safe dose in humans. After trialling some variant forms earlier in the year, on 17 December 1951, the first patients got Rimifon itself – only those where all other treatment options had failed. After just a few days, most showed signs of recovery. The results were so incredible that when they were presented to Roche's CEO, he reportedly said 'we should not worry about profits'. Instead, he wanted Roche to commit to making the drug available to everyone who needed it, at a price they could afford.

Roche was certain it had a fairytale on its hands. But other companies were also working on new antibiotics for tuberculosis. Across New York Harbor from Staten Island, a doctor called Walsh McDermott had been supplied with a new antibiotic for his most hopeless cases by scientists at Squibb, another firm with research headquarters in New Jersey. McDermott himself suffered from tuberculosis and had received the antibiotic after a recent surgery. Now in late 1951 he was giving it to his patients. And across the Atlantic in Germany, a research team at Bayer led by Gerhard Domagk, who had discovered Prontosil, was trialling yet another new antibiotic.

Three new antibiotics from three pharmaceutical firms. Yet remarkably, they were all the same: a molecule called isoniazid. Kogel's press conference was only the latest twist in what was turning out to be something of a corporate nightmare.

On New Year's Eve 1951, Roche scientists working on isoniazid had heard a rumour that scientists at Squibb – just half an hour away in New Jersey – had been testing exactly the same substance. A hastily arranged meeting in mid-January with representatives from both companies confirmed it. In what appears to have been a genuine coincidence, they had somehow found the same drug. It was uncannily like the story of Prontosil: isoniazid was an intermediate molecule in the steps of reactions chemists were using to make drug candidates, but had turned out to be more effective than the candidates themselves.

Isoniazid had already been manufactured in 1912 by scientists in Prague. Both Roche and Squibb knew that it was therefore not patentable as a new substance. They came up with a plan for a staggered announcement of the news of their discovery in April 1952. First, they would unveil the results jointly at a public symposium of physicians in New York, only informing the press on the following day. The situation was tense. Was it really a coincidence that both had been working on the same drug? The possibility of industrial espionage was surely never far from anybody's mind, but both parties tried to be amicable. When the Squibb CEO visited the Roche researchers, he made sure to praise their fair play.

Kogel's surprise press conference on 20 February had now torpedoed this careful plan. In its aftermath, neither Squibb nor Roche controlled the narrative. The journalists picked up the good news story and ran with it. In the *New York Times*, a journalist described Lazarus-like transformations, as mortally ill patients got up from their beds and walked. The doctors at Sea View described the effects as beyond anything they had ever seen. Meanwhile the *Washington Post* speculated whether it was finally 'KO for TB?'.

In Germany, the mood was less ecstatic. The Bayer CEO was compelled to depart from what he referred to as Bayer's 'customary reserve' when he told a radio interviewer that these American drugs were indistinguishable from the drugs Bayer had been developing in secret. There was patriotic outrage at not one but *two* companies based in America stealing Bayer's thunder. One German magazine even claimed it was another example of wartime plunder: apparently, during WWII, occupying forces had stolen research data from locked safes and passed it on to American companies.

Knowledge had certainly been transferred by WWII, but it hadn't needed the looting of safes. Robert Schnitzer, one of the Roche scientists who had discovered the activity of isoniazid, was Jewish. A citizen of Germany, he had worked for Hoechst in the 1930s (within IG Farben, Bayer's parent company) researching tuberculosis, until he was stripped of his job in 1938 and sent to the Buchenwald concentration camp. Promising to leave the country in order to be released in January 1939, he fled to France and then on to America. By the end of the war, he worked for Roche – Bayer's rivals – in New Jersey.

As a European firm headquartered in Basel, Roche wanted to remain on good terms with Bayer. They agreed a joint statement that was published across eight different journals in August 1952, declaring that both companies had independently discovered isoniazid.

The convergence of three different firms on the same synthetic molecule shows how antibiotic research was becoming more informed – or less scattershot. The companies weren't independent: their convergence could be explained by the dissemination of knowledge through scientific journals and conferences. In 1945, a French scientist working in Paris had shown that a molecule called nicotinamide, a form of vitamin B3, could treat tuberculosis in guinea pigs. However, the effect was far too weak: a human would need to take over 150 grams of pure vitamin daily, making it far too expensive as a treatment. Three companies had been looking among derivatives of vitamin B3 to find one with greater efficacy. Using the standardised tricks of the trade for manipulating molecules, they all discovered isoniazid. As a Roche-sanctioned history of the events put it, there was 'an inevitability to the simultaneity'.

Beyond the simultaneity, there was something more to be explained. At the point of realising that the new molecule under development could not be patented, why did the companies not stop their work? True, profits could be made without a patent.

But the personal experience of the researchers was also an important contributor. In Germany, Domagk had looked at the post-war tuberculosis situation with horror, reflecting that the disease threatened Germany 'to a degree that we have never previously known and could not have imagined'. In America, four-fifths of everyone who died of an infectious disease between the ages of fifteen and thirty died of tuberculosis. Everybody knew somebody with the disease; everybody knew the way it sapped life out of the young and returned to carry the old away. It was unthinkable that a promising antibiotic would be abandoned.

Without the option of a patent, the three companies all knew that the way to make money from isoniazid was to secure their branded form as *the* form in the public's mind. Domagk wrote in December 1951 to his superior that the only way to take over the global market was to produce enough isoniazid to meet all requirements. The manufacturing race was on.

In the US, the power to approve a new medicine lay with the Food and Drug Administration (FDA). Scientists at Roche and Squibb were jockeying for position to get their respective products approved, which would be available only on prescription. In Britain, clinical trials had started in March 1952, the names of the centres censored from newspapers because so many people were desperate to be treated. Because isoniazid was synthetic like Prontosil, rather than brewed up by microorganisms like penicillin or streptomycin, it could be easily manufactured by simple chemical steps. Those steps were well within the reach of many companies.

By a quirk of British law, isoniazid also fell into a strange legal loophole. It was not yet registered as a medicine, so wasn't under the remit of the Ministry of Health (unlike, for example, penicillin). Yet on the other hand, because no good evidence yet existed of its toxicity, it also wasn't on the poisons list that restricted the sale of dangerous chemicals. Legally speaking, isoniazid was 'just' a chemical – a chemical that could be purchased without prescription from anybody who chose to sell it. Within two months of Kogel's press conference, isoniazid was being manufactured by several firms in England under a bamboozling variety of names: Rimifon was joined by Nydrazid, Pycazid, Mybasan and others.

There were immediate fears for the development of antibiotic resistance. 'Abuse of TB drug feared in Britain', warned the *New York Times*: 'injudicious use' could produce 'a super-resistant strain'. The *Manchester Guardian* reported that medical authorities were concerned about the dangers of 'indiscriminate manufacture'. Isoniazid had been heralded as better than all existing treatments, and authorities worried that it was being smuggled into sanatoria and used to treat patients under the noses of doctors.

The concerns about resistance were well founded. While the *New York Times* was warning about resistance, in Germany, clinicians were isolating isoniazid-resistant strains of bacteria and sending them in the post to Domagk at Bayer. Sceptical observers urged against celebrating the early successes at Sea View as definitive. They knew from experience that resistance could jeopardise even the most impressive antibiotics.

Isoniazid had arrived just as René Dubos, whose insight had set Waksman on the path to streptomycin, was finishing the proofs of an epic history of tuberculosis. It was a disease that had touched Dubos's life. In 1934 in New York, he'd married his wife Marie Louise, a talented pianist who had suffered from tuberculosis as a child and also, like him, suffered from rheumatic fever. At his request, the couple received their final wedding blessing not in front of the church's altar but behind it, in a replica of the shrine at Lourdes. It didn't work. In 1938, Marie Louise had begun to cough up blood on a road trip through the American West. Four years later, after spending time in a sanatorium, she was walking down the street near Carnegie Hall when she realised in a panic that she would never play the piano again. Within a month, she was dead. In an elegy for her, Dubos wrote that his wife had 'wanted only to know the springtime of life'.

Dubos had subsequently remarried, but he hadn't left tuberculosis behind. His second wife Jean was as obsessed as him, if not more so. She had been a researcher on tuberculosis in his laboratory before they married, and afterwards spent years in libraries researching the history of the disease. The result, co-written with her husband, was a biography of tuberculosis: *The White Plague*. The couple wrote most of it in their country house outside New York, with Marie Louise's grand piano standing like a *memento mori* in the living room. René and Jean were barely interested in antibiotic treatment, taking a much more historical and sociological perspective on the disease. In the small section where they discussed antibiotics, they noted that resistance to streptomycin and PAS seemed inevitable; despite the dramatic effects the new drugs could have initially, after weeks and months the disease could creep back. As their book went to the printers, news of isoniazid made them add a rushed sentence expressing their hope that the good news was real. But then, at the last minute – apparently too late to alter the main text – they changed their mind again and added a cautionary footnote. They warned that isoniazid, for all the 'exciting headlines and photographs . . . will probably be regarded as just another treatment when re-evaluated in the light of experienced judgement'. The couple whose lives had been shaped by tuberculosis believed that its eradication from society was unlikely to be solved by antibiotics.

McDermott, the doctor who had received isoniazid for his own tuberculosis, was more optimistic. The experience of working in medicine before the sulfa drugs were introduced had allowed him to witness what he called 'the introduction of a new era'. As a young doctor, he'd been among those who had seen one of the first American treatments with Prontosil: a woman who worked for a German export firm in New York had contracted apparently fatal blood poisoning, but at the emphatic request of her family, she was treated with 'a German dye', to the bemusement of the American medical team. She recovered. At the time, the doctors had all dismissed this as a coincidence. But soon, after the widespread introduction of the sulfa drugs, they realised they'd witnessed a moment of history.

The year that he'd finished medical school, McDermott had been diagnosed with tuberculosis. He was sent away to a sanatorium where he spent seven months convalescing. In total he would make nine trips to hospital over the next nineteen years. His career progressed alongside his illness, and he eventually became Director of Infectious Diseases at the Cornell university hospital in New York. Despite this, he still wasn't cured. The *M. tuberculosis* inside his lungs would occasionally burst out and leak into new areas, causing flare-ups of acute disease. McDermott insisted on continuing to work even as a patient, directing the research efforts of his division from his bed.

After he'd experienced the benefits of isoniazid himself, McDermott wanted to test it further afield. Everyone who had been given the antibiotic so far had received it as a last-ditch effort: they had already been given most if not all of the best-known treatments, such as streptomycin and PAS. Could giving isoniazid first produce the same amazing effects? That needed to be tested. However, McDermott thought it would be unethical, given the established benefits of those other antibiotics, to withhold them and instead only give patients isoniazid. It was unconscionable to risk a patient's recovery for the sake of scientific experimentation. Giving isoniazid alone would only meet his ethical standards if it were impossible to give streptomycin. Since streptomycin was usually administered with daily injections to sanatorium patients, that meant finding a region of America with high levels of tuberculosis but essentially no sanatoria.

McDermott found the perfect place. He decided to trial isoniazid in collaboration with people who lived over 2,000 miles away, in a remote rural community called Many Farms in the state of Arizona. In their own language, they called themselves the Diné. McDermott knew them as the Navajo.

In the early 1950s, the Navajo lived in what McDermott described as 'regal poverty'. Their homes were mostly small huts called hogans: single-room constructions with heaped walls of logs and earth, without windows, where up to fifteen people might live together. These were well insulated against the extreme temperatures of the high mesa: anything from −29 °C in winter to 43 °C in summer. But, like the cramped tenements of Glasgow or the slum housing of Manhattan, the windowless hogan was also ideal for the spread of a respiratory disease like tuberculosis.

Many Farms was a fairly typical community within the reservation governed by the Navajo Nation. The flat and arid valley floor, about 5,500 feet in elevation, was bordered by rough badlands to the east. Walking over to the west you'd climb the valley side into higher ground, sage brush and sparse grass giving way to juniper-clad uplands until you reached the base of the mesa known as the Black Mountain. About 2,000 people lived there, scattered across 800 square miles in small collections of hogans that never reached the critical mass of villages, connected only by often-impassable dirt roads. The nearest hospital was over fifty miles away. Infant mortality was up to three times higher than in the rest of the US. The Navajo had been exposed to the 'modern' medicine of the white doctors since at least the 1880s when physicians had begun to appear in their reservation, but before the antibiotic era, there was little to choose between their remedies and those of the traditional Navajo medicine men. That was most marked for infectious diseases, where neither the Navajo nor white physicians had many effective remedies. But when antibiotics arrived, the Navajo quickly recognised their benefits, particularly for tuberculosis. Here was a medicine that clearly worked.

Early trials of isoniazid on the reservation had impressed both the doctors and the Navajo. After discussion with McDermott and others, in April 1952 the Navajo Tribal Council passed a resolution to donate $10,000 to the Department of Medicine at Cornell, to assist them in carrying on their work as consultants to the use of isoniazid on the reservation. The results from the reservation were rapidly cited: amid the uncertainty and chaos of the isoniazid situation, in July 1952, the British government's committee on tuberculosis treatment asked to see data from the Navajo reservation.

The work went well. One Navajo leader, Billie Bicenti, stood up at a crowded meeting of the Tribal Council and reflected on the experience. 'We knew we were afflicted, but never knew to an extent what was ailing us,' he said. 'Now these people have made it possible to see and realize that there is something that can be done about those things that have been harming us and killing us right along'. The work was a collaboration where the academics and the Navajo tried to talk to each other as equals; beyond supplying the medication, drug companies weren't involved. Navajo representatives went to New York to see the Cornell laboratories. The Chairman of the Tribe, Sam Akeah, told the Tribal Council that he had been impressed with the research into tuberculosis and other diseases, but bemused by New York. 'It remains a mystery,' he said, 'as to why people congregate to live in a place like that'.

McDermott wanted to set up a longer-term trial to test whether bringing the very best medical care, represented by Cornell physicians, could improve the Navajo's general health as dramatically as isoniazid had handled their tuberculosis. After enrolling anthropologists and linguists as well as doctors, they took the proposition to the Tribal Council and selected Many Farms as the trial's focus. In a room in September 1955 crowded with men, women and babies – the local irrigation office was the only place big enough – the leader Annie Wauneka explained the proposals to her fellow Navajo. Wearing traditional dress of a blanket over her shoulders, turquoise beads and a long skirt, she spoke of the improvements in tuberculosis treatment that the past few years had seen. A clinic with X-ray equipment and medicine would be brought into the heart of Many Farms: 'We members of the Tribal Council have discussed this clinic; we have considered all sides of the question and we have concluded this is one of the best things being done for our people.'

The Many Farms experiment brought a system of primary care to the community. Navajo medicine men, many of whom had experienced tuberculosis themselves, became health visitors. The Navajo language was so unfamiliar to outsiders that Navajo men served as 'code talkers' in the US Army in WWII: they could encode, transmit and decode messages far faster than any machine. The code talkers had created new words for military terms; now, the medicine men needed to do the same for science. Dissecting a sheep to try and establish a translation map for human anatomy revealed that the Navajo had no word for 'lungs' as distinct from the combined organs of the upper chest. There were other subtleties too. Doctors treating tuberculosis often wanted to know whether patients had seen blood in their spit. But translating the apparently simple question 'What colour is your spit?' literally into Navajo was impossible, because the Navajo language has no abstract concept of 'colour'. A better way to frame the question would be: 'Is your spit yellow or red?'.

Despite these difficulties, the health visitors and the doctors established a cordial relationship. McDermott's own experience as both doctor and patient with tuberculosis was mirrored in theirs – the divisions between doctor and patient blurred.

Since tuberculosis is a slow-burning disease, five years was too short to measure the effects of isoniazid at Many Farms. But the researchers could investigate transmission, which they did by look-ing at children who'd been born during the study. A skin test for an immunological reaction for tuberculin, the purified product of *M. tuberculosis* cells that Robert Koch had once touted as a treatment, would come up positive if a person had previously been exposed to the disease. At the start of the study period, over 30 per cent of five-year-olds starting school tested positive. At the end, that had fallen to only 5 per cent. Tuberculosis transmission in the cramped hogans had been drastically reduced. Other infectious diseases showed less impact. For example, the incidence of diarrhoea didn't change, because the presence of doctors didn't fix the lack of run-ning water. But childhood ear infections – treatable with penicillin or sulfa drugs – did reduce in the last year of the study. The two single proven improvements to health, as McDermott and his col-leagues noted with a wry humour, 'did not actually require the presence of a physician'. They were due solely to antibiotics.

McDermott was keenly aware that antibiotics were the main part of the 'modern' medicine that the Navajo admired. He began to turn a more anthropological eye to the system of medicine that he had been trained in. To him, it seemed there was a tension between the paradigm of the patient-physician interaction – a personal encounter – and new drugs like antibiotics, where that encounter was just a way of getting the drug. Because seeing a doctor was *the* delivery route for medical care, antibiotics and other new technologies had been engrafted onto a centuries-old system. Within a few short decades, the doctor's role had become more like that of a pharmacist. McDermott felt that these new scientific developments were stressing the medical profession 'almost to the bursting point'. Medicine was running away with itself.

In *The White Plague*, René and Jean Dubos had pessimistically speculated that isoniazid would do nothing for the transmission of tuberculosis. The trial at Many Farms had shown that they were wrong. Dubos himself had been invited by McDermott to join the studies in 1953, which he did despite the difficulty of visiting Many Farms. (At one point, McDermott and Dubos had to push their plane out of sand on the single runway airfield before it could attempt to take off.) McDermott was strongly influenced by Dubos's ideas about the social character of disease. Many Farms showed that even the best medical care was incapable of improving the general health of people who lived in extreme poverty. Antibiotics were a powerful tool, but they couldn't fix every social ill on their own. Good food, clean water and sanitation were the bedrock on which medicine needed to stand. But the careful assessment of those like McDermott and Dubos was being drowned out by the triumphant cacophony erupting from pharmaceutical companies, flush with their success.

One of the side-effects that the doctors had noticed at Sea View was euphoria. It gradually became clear that some of the most dramatic effects which had been taken as indicative of health – consuming three bowls of cereal, dancing in the corridors – were not due to any actual change in the condition of the tuberculosis. The patient could *feel* better, even when X-rays showed their condition hadn't changed. Isoniazid was a long way from Paul Ehrlich's vision of a magic bullet that left human cells unscathed – it was somehow affecting the human brain as well.

In Glasgow, one doctor was inspired by the reports of euphoria to immediately start using isoniazid on patients with severe schizophrenia. Most of his patients had previously failed to respond to electroconvulsive therapy, where the patient is violently shocked with electric current to fry their brain, but remarkably, 45 per cent of them improved with isoniazid. One fifty-year-old man had believed that he was being controlled by atomic energy directed by 'the power behind the throne'. After isoniazid treatment, he became noticeably brighter and more alert. When asked whether he was still being controlled by atomic energy, 'he would no longer admit to the delusions noted previously'. How a tuberculosis drug could make a man disavow his paranoid delusions was a mystery, but it had to be a *chemical* process. The doctor concluded that the dramatic effect of isoniazid offered support for 'theories of a disturbed cerebral biochemistry'.

By 1957, US doctors had announced that a close molecular relative of isoniazid called iproniazid seemed to offer a powerful therapy against depression. Iproniazid was less effective against tuberculosis but had greater psychic effects, producing feelings of vitality and wellbeing. Medicine didn't have the language for this new type of drug. Struggling for words, the best the *New York Times* could do was call it a 'psychic energizer' – within a few years, the term would be antidepressant.

The discovery of antidepressants was a paradigm shift in psychiatric illness. Before isoniazid and iproniazid, the best treatment for those with severe depression was tranquilisers, which aimed to sedate and numb the patient rather than change their state of mind. In 1957, the market in tranquilisers was worth $195 million a year. Now, it seemed there was something even better. Researchers gradually uncovered a plausible chemical mechanism: iproniazid blocked an enzyme in the brain that degrades serotonin, producing higher serotonin levels. Thus was born the 'chemical imbalance' theory of depression. Some forms of depression seemed to be alleviated by a drug that increased serotonin levels. Therefore (went the theory) the depression must have been *caused* by low serotonin levels. The reasoning was shaky, but in this framing, a pharmaceutical product could 'top up' the chemical balance. What's more, iproniazid was a generic drug, cheap and plentiful because of the ramped-up production for tuberculosis treatment. In this new chemical utopia, it seemed possible that asylums would go the way of sanatoria, rendered obsolete by the very same medicines.

By 1958, around 400,000 Americans were taking iproniazid, supplied by companies including Roche. One doctor who had been a pioneer of iproniazid was dismayed at the widespread use. It was a toxic drug that needed a patient to be kept under close observation, he warned, 'not a "pep pill"' to be sold like a refreshing caffeinated drink'. Iproniazid was toxic to the liver, and in a high-profile case, the drug was blamed for a woman's death in California. The drug was eventually withdrawn, but the growth of the antidepressant industry continued.

The discovery of isoniazid heralded a new era in the treatment of tuberculosis: truly effective antibiotic therapy. When doctors at Sea View had first compared isoniazid against the existing treatments of streptomycin and PAS, they felt that it was fruitless to look for the single best antibiotic. 'It seems safe to assume', they wrote, 'that an assault on the tubercule bacillus from all quarters is desirable'. So it proved. The triple combination of streptomycin, PAS and isoniazid was a winning antibiotic cocktail. Isoniazid was the bactericidal component that had been needed to effect a transformation, and the combination of three drugs made the emergence of triple resistance statistically improbable. After the introduction of isoniazid, many patients could now recover from pulmonary tuberculosis with antibiotics alone. Invasive surgeries like thoracoplasties became redundant; my grandfather Philip's scars remained etched on his body, a reminder of a more brutal age. Today, the first-line treatment against tuberculosis is a cocktail of four antibiotics, refined through successive clinical trials like the one at Many Farms. All were developed before 1966. Isoniazid is one of them.

Within a decade of isoniazid's discovery, sanatoria across the Western world had emptied: the one at Grey Towers, where my grandfather stayed as a young man, became little more than a haunted house for local teenagers to explore; the last tuberculosis patients left Sea View in 1961. The leafy forest that had provided a pleasing vista from the elegant pavilions began to encroach, winding its green creepers through the decaying bricks, as saplings took root in rooms that had once been filled with patients. Without preservation, the memories of tuberculosis decayed too. It became thought of by many people as a nineteenth-century affliction of pale poets and dying composers – part of humanity's past, not its future. But that belief was dangerously naive. As Jean and René Dubos had warned, the mere existence of antibiotics would not be enough to eliminate a disease that was inextricable from social inequality.

Yet at the time, the miracle of isoniazid seemed set to deliver humanity from an ancient foe. Isoniazid was a molecule of multiplicities. Discovered three times, it had a double effect, impacting both bacteria and the human brain. Its powerful psychological effect on patients mirrored the wider atmosphere created by antibiotics. There was a giddy euphoria in the air. If tuberculosis and depression could be conquered by the same molecule, what else might yield to the power of the pharmaceutical industry? Once doctors had championed the management of disease through social reform, exercise and diet, but now it seemed as if there really might be a pill for everything.

Age of excess
Tetracycline

It is assumed that we are ill and are made well, whereas it is nearer
the truth to say that we are well and are made ill.

Thomas McKeown

It was 1950, and the pharmaceutical executive John L. Smith was
dying of lung cancer at the age of sixty-one. Even on his deathbed,
he was worrying about the fate of his company, which he'd joined as
a seventeen-year-old: Chas. Pfizer & Co, otherwise known as Pfizer.
Smith had been succeeded by his one-time protégé John McKeen,
but he was still kept informed of Pfizer's latest breakthroughs. The
two men were close. In 1943, they had kept anxious vigil over a teen-
age girl who was dying of fever. After a desperate appeal from her
father, penicillin made by Pfizer had been flown to her hospital in an
Army B-24 Liberator bomber. The two executives had watched her
progress, amazed as the penicillin delivered her from the infection
and saved her life. Now Smith himself was dying.

Yet that wasn't what was on Smith's mind.

Following streptomycin's discovery in soil, Pfizer had recently
started its own massive soil-screening programme, and Smith had
just heard that a promising new antibiotic had been found. But he
was worried. 'Don't make the mistake we made with penicillin and
hand it over to other companies,' Smith warned McKeen. 'Let's sell
it ourselves. Go into the pharmaceutical business if we have to.'

Today, Pfizer is one of the world's biggest pharmaceutical
companies – the first ever to bring in a global revenue of $100 billion.

But as Smith lay on his deathbed, it wasn't. Pfizer's history was as a wholesale supplier of chemicals, manufacturing them and selling them on to other companies. At the turn of the twentieth century, Pfizer's mainstay had been citric acid, a harmless chemical additive with a tart taste, used in everything from paper manufacture to carbonated soft drinks. Making it required thousands of gallons of imported juice from lemons, limes and oranges grown in Southern Europe. In particular, Pfizer needed Italian lemons, sharpened under the Mediterranean sun then shipped over the Atlantic to its factory in Brooklyn. But lemon producers in Italy formed a powerful cartel under the control of the Mafia – some academics have even argued that the market for lemons was the key factor that led to the Mafia's formation in Sicily in the 1880s. The Italian cartel could unpredictably squeeze the supply, damaging Pfizer's business.

Pfizer found such a lemon grab unacceptable. So in 1919, they started trying to free themselves, researching a method to mass produce citric acid from American sugar. Smith came to know the process intimately: an unappealing black fungal mould, *Aspergillus niger*, was grown in huge vats, fed with corn syrup and belching out citric acid. Smith had overseen the first citric acid fermentation plant himself, the stench of the stinking slops rising over Brooklyn. By 1924, after Prohibition had led to a rise in soft drink consumption, millions of Americans were happily drinking citric acid that had never seen the inside of a lemon. The citrus taste gave away nothing about its origins.

Pfizer was a paranoid company. It chose not to patent its fermentation method, since that would have meant publishing too many details. Instead, Smith kept the method an industrial secret like Willy Wonka in his chocolate factory. But in WWII, when the US government had appealed for companies to help with penicillin production, Pfizer had participated in the programme. In the spirit of wartime cooperation, they had shared their secret fermentation expertise with others, transforming the amount of penicillin that could be made. That decision had undoubtedly saved thousands of lives by enabling mass production. But from his deathbed, Smith was warning McKeen not to make that 'mistake' again.

Cooperation had destroyed Pfizer's competitive advantage. Pharmaceutical companies that had previously bought penicillin from Pfizer soon started making their own, and surplus penicillin started to pile up in Pfizer's headquarters. In just over half a decade after 1945, penicillin's value had fallen by 99 per cent. They'd experienced similar results with streptomycin. McKeen himself had told other companies, 'if you want to lose your shirt in a hurry, start making penicillin and streptomycin'. But when Smith had argued that Pfizer should sell streptomycin as a medicine themselves, the Vice President of Sales had opposed him. For Smith, being a manufacturer in a crowded market was no good. The dream was to have independent control of an exclusive molecule, woven into existence from *your* microbes: millions of lemon farms in a vat.

Both penicillin and streptomycin were made, like citric acid, by microbes. However, their initial discoveries couldn't have been more different. Pasteur's dictum that 'chance favours only the prepared mind' fitted Fleming's serendipitous discovery of penicillin perfectly: an apparently trifling observation that needed a great mind to interpret it. Of course, Fleming had only identified the strange behaviour of a mould juice, and the actual identification of the active ingredient, its purification and manufacture had taken thousands of people working in collaboration, including at Pfizer.

In Waksman's version of the streptomycin story, the antibiotic had come from a deliberate and targeted search. This method had little of the romance of serendipitous scientific discovery. Waksman's claim was that his methods served to *eliminate* chance, assigning a fixed way of working that any trained scientist could carry out. The truth that the discovery owed a great deal to Albert Schatz's insight and diligence had been deliberately obscured. As with penicillin, though Merck had technically had the right to assert exclusive rights over the new antibiotic, the same wartime environment with government pressure to upscale production had resulted in streptomycin being licensed to many companies. Waksman's team found other new antibiotics, patenting as they went, but none was as significant as streptomycin.

Fleming himself was unconvinced by systematic approaches. In 1946, shortly after he had received the Nobel Prize, he warned a meeting of British microbiologists not to get too excited about Waksman's methods. 'This plodding along appeals to some,' he told them, 'but no one can expect the results to be revolutionary'. He was wrong. The antibiotic gold rush was about to begin – and in the process, it would reshape the pharmaceutical industry.

The penicillin 'mistake' had left Pfizer with huge manufacturing capabilities for antibiotics: now they just needed a monopoly. The Patent Office's ruling on streptomycin in 1948 had set a precedent that natural molecules produced by microbes could be patented. At that time, a patent guaranteed a seventeen-year period of exclusivity, during which the patent-holder could prevent other companies making or selling the antibiotic. Predictably, the other companies that had invested heavily in the production of penicillin and streptomycin were thinking similarly to Pfizer. A crazed search began for the most obscure soil possible: the more obscure, the more likely it might contain something that competitors wouldn't find in their own samples.

This strange collective mania took hold across the pharmaceutical industry. Parke-Davis sent their entire sales division off with plastic bags to bring samples back from their travels. Bristol-Myers Squibb sent their annual report out to shareholders with an envelope for them to send a teaspoon of soil back in. In a partnership that fused God and Mammon, Eli Lilly arranged with the Christian and Missionary Alliance for its missionaries in places like the jungles of Borneo to collect soil and send it back. The commercial desire for exotic soil also aligned with other interests, as when a report from the British Colonial Research Council included an update on the continuing search for antibiotics in Malayan soil.

If a company discovered a unique antibiotic that killed bacteria and was not obviously harmful to patients, the commercial potential was huge. As antibiotics proliferated, so too did a sense that these once life-saving medicines should also be applied to even minor ailments like earaches or stomach troubles. Higher sales of the antibiotic would mean more profits. As long as the substance was protected by a patent, those profits would not evaporate like they had done for penicillin and streptomycin, but continue for up to seventeen years. Almost every aspect of the American healthcare system, from prescriptions to drug regulation, was unprepared for the antibiotic avalanche that was coming.

Pfizer's own soil screening programme took in 135,000 soil samples, and its scientists conducted 20 million tests. One Pfizer scientist described the breadth of their search:

> We got soil samples from cemeteries; we had balloons up in the air collecting soil samples that were wind borne; we got soil from the bottoms of mine shafts; we got soil from the bottom of the ocean; we got soil from the desert; we got it from the tops of mountains and the bottoms of mountains and in between.

The programme was motivated not by medical need but by commercial concern: Pfizer wanted its own patented antibiotic to sell to patients. Ironically, despite all this far-flung collecting, the most promising antibiotic that Pfizer's scientists had identified came from an agricultural field in the state of Indiana. That new antibiotic killed a wide range of bacteria when the scientists tested it in the laboratory. As Pfizer president, John McKeen was given the choice of the name for the new product. He dubbed it Terramycin, justifying his choice by saying that he 'wanted a name connected with the earth . . . because it came from the earth'.

Smith, McKeen's mentor, died in July 1950, leaving McKeen to take Terramycin forward with the warning ringing in his ears. Smith's remedy for Pfizer's woes had long been that Pfizer should start selling antibiotics directly to doctors themselves. Others at the company now agreed, and plans were afoot for Terramycin to become Pfizer's flagship product.

There was just one problem: Pfizer was not a pharmaceutical company, and had no experience in that area. So it had enlisted an expert who did. Pfizer was about to launch one of the most aggressive marketing campaigns the medical world had ever seen.

The New York advertising firm William Douglas McAdams specialised in medical products, and in 1942, they hired a young doctor called Arthur Sackler. The eldest of three brothers, Sackler already had a full-time day job as a doctor in a psychiatric hospital, but he was driven by an almost inhuman energy that saw him working nights and weekends in the firm's offices, where he combined his medical knowledge with a prodigious gift for writing advertising copy. The firm's tactics were to target not the patients who used the drugs, but the doctors who prescribed them. When Pfizer contracted them to market Terramycin, it was Sackler's responsibility to come up with the campaign.

Sackler saw that to sell this medication, he needed to first turn Pfizer into a pharmaceutical company. The main problem to overcome was that because Pfizer was a manufacturer of raw materials, most doctors would have never heard of it. In advertising parlance, Pfizer had no brand recognition at all. So, the first advertisement he placed in a medical journal didn't even mention the new antibiotic by name. It consisted only of the words 'Terra bona' – Latin for 'the good earth' – with the Pfizer logo. He was seeding the ground for a momentous announcement, encouraging speculation about what Pfizer had in store without even mentioning the product itself. It was the first time a 'teaser ad' had been used in pharmaceutical advertising.

Then, Sackler upped the ante. The ads for Terramycin were a masterclass in hyperbole. Terramycin, they boasted, had 'one of the most complex structures ever to be found in nature'. (It didn't.) Terramycin was presented less as a drug, and more like a remedy plucked from the garden of Eden. It was, the adverts suggested, a cure-all. Whereas research for an antibiotic like streptomycin had quickly focused on its effect against tuberculosis – a specific disease – many of the newer antibiotics were not developed with particular illnesses in mind: companies just wanted something that killed as many bacteria as possible. In June 1950, one scientist talking about this difference had spoken of the 'broad spectrum of activity' of the new generation of antibiotics. Sackler seized on the phrasing. When the first ad that mentioned Terramycin by name appeared the next month, the copy proclaimed it as a 'broad-spectrum' antibiotic.

The term had never before appeared in the medical literature. It suggested that Pfizer's product was something revolutionary. A doctor prescribing a broad-spectrum antibiotic could remain ignorant about what was causing the patient's infection. If a doctor prescribed penicillin and the species causing the infection was Gram-negative rather than Gram-positive, it would be useless. And since doctors could rarely wait for a microbiological test to identify the species causing an infection, a broad-spectrum antibiotic provided a safer bet. Doctors were being implicitly encouraged to not think too much about the underlying causes of the disease in the patient, but rather to respond with a kneejerk reaction: a prescription for Pfizer's antibiotic, Terramycin. Doctors were to become pill-pushers.

Terramycin was not the first broad-spectrum antibiotic – streptomycin also fit that description. But deploying the phrase was a stroke of advertising genius. The campaign for Terramycin influenced the idea of how antibiotics ought to be prescribed: a broad-spectrum antibiotic became a doctor's trusted friend. After all, wasn't it better to be safe than sorry? The division of antibiotics into *narrow* and *broad* has remained: medical students in universities around the world are still taught about antibiotics using these terms, which owe their initial popularity to Arthur Sackler.

The triumphant tone of Sackler's ads made it sound like Terramycin represented a huge medical advance. As Pfizer's scientists had established, that was false. In fact, Pfizer knew that Terramycin had near-identical effects to another antibiotic that had been discovered two years previously by Lederle, a rival company. That antibiotic was called Aureomycin, named after the golden colour of the bacterium that produced it, another *Streptomyces*, this time *Streptomyces aureofaciens* (the 'gold-maker'). Terramycin was so similar that some Pfizer researchers initially mistook it for Lederle's antibiotic.

Under US patent law, Terramycin could still be patented if it was a different molecule from Aureomycin. But to prove the difference, Pfizer needed to be able to describe its structure – which its scientists didn't currently know. So McKeen enlisted the chemist Robert Burns Woodward as a consultant, asking him to work out what he could. Woodward apparently wrote down every known fact about Terramycin on a large piece of cardboard and promptly reasoned out its entire structure.

Woodward's conclusion was concerning. In molecular terms, Terramycin and Aureomycin were like identical twins, so similar that only their nearest and dearest would be able to tell them apart. They shared over fifty atoms in common arranged into a four-ring (tetra-cyclic) structure that differed at only two molecular positions. Both were minorly adorned versions of a central molecule, tetracycline, which alone had the same antibiotic effect. It was an echo of the situation with the sulfa drugs. The similarity between these two related tetracyclines, as molecules in the chemical family became known, was not a coincidence. Inevitably, companies using similar soil-screening methods were discovering similar antibiotics.

The good news was that the presence of those minute differences meant that legally the molecules were different substances, and so under US law each could be patented independently. Pfizer's patent had been filed in November 1949. The company's market researchers estimated that antibiotics would account for $1 of every $4 spent on prescription drugs in 1950 – and they wanted their piece of that growing pie.

Pfizer's transition into a pharmaceutical company had been swift. The paper announcing Terramycin had been published in January 1950. In less than two months, the drug was approved in the US by the FDA. In the admiring words of *Business Week*, Pfizer had set a 'speed record' for moving a product from the laboratory to clinical use: the whole process from discovery to use had taken less than a year. Within an hour of the drug's FDA approval, eight salespeople – Pfizer's nascent sales force – began to telephone 'a hundred wholesalers each, offering a very steep introductory discount'. Then from June 1950, Sackler's advertising campaign blitzed the medical profession like never before. In the first two years of marketing Terramycin, Pfizer would spend twice what they'd spent developing it. Their sales force soon swelled to hundreds of people. Fellow medical advertisers thought that the campaign had changed the marketing model of drugs forever. In the words of one of Sackler's employees, when it came to pharmaceutical advertising, he 'invented the wheel'.

Sackler burnished his own reputation alongside Pfizer's, but the long-term repercussions would be even greater. Keen to sell his own drugs rather than simply advertise those of others, he and his brothers bought the pharmaceutical firm Purdue Pharma in 1952 and continued expanding their empire. In 1995, eight years after Arthur Sackler's death, the family-owned company began marketing OxyContin, a powerful opioid painkiller. The advertising campaign for OxyContin targeted doctors, extolling the virtues of the drug and claiming that it was non-addictive. In fact, Oxy-Contin was highly addictive. But even as evidence of its harms accumulated, the Sackler family amassed a huge fortune from the patented drug. In the twenty-five years after OxyContin's introduction, the epidemic of opioid addiction and overuse in which it played a major role killed over 450,000 Americans. There was more than a family resemblance between the advertising campaign for OxyContin and the tradition of exaggeration that Sackler had established with antibiotics. Drugs did not simply serve medical need – they could create it.

Sackler's early work in advertising Terramycin was part of a powerful post-war shift in pharmaceutical advertising. Companies began to target not the patients who took the drugs, but the doctors who prescribed them. From 1952, Pfizer purchased more than two-thirds of all antibiotic advertising in the *Journal of the American Medical Association*. With Terramycin, Sackler helped to forge a narrative for the whole pharmaceutical industry: companies were benevolent life-givers, for whom profit was a happy side-effect.

Advertising shapes the world, and it shaped how doctors and scientists thought about antibiotics. Tracing that history now is challenging, not least because although the medical journals of the day were packed full of adverts, librarians tore them out before archiving them. But like dark matter, their influence can still be detected, a powerful force that framed how doctors thought about the new antibiotics. By focusing endlessly on their natural origins and their links to the 'good earth', companies obscured the potential problems their mass use might cause. Antibiotics that had once been valued more than gold and reserved for the most serious illnesses were now being given indiscriminately, producing an increase in the proportion of resistant strains. This meant that when patients did fall seriously ill, doctors increasingly discovered that the treatments expected to save them were ineffective, because of the widespread usage for trivial ailments. In December 1950, the American doctor Max Finland warned against the dangers of antibiotic proliferation. After all, he asked, where did such nonsensical logic end? From the moment a child was born,

> is not the infant thereafter to receive some antibiotic each time it sniffles, or looks poorly or has a fever? And are not all of his family to be treated to assure his being freed of sources of infection? And is he not to brush his teeth with an antibiotic-containing dentifrice twice daily as long as he lives or has teeth? And so it goes on until his dying day!

Finland's satire was not far from the truth. Throughout the 1950s, antibiotics found their way into almost every product imaginable. Readers of *Good Housekeeping* were advertised nose sprays, mouthwash, toothpaste and shaving balms – all containing antibiotics. An antibiotic made by Pfizer was even put into the exploding harpoons of whalers as an 'on-location meat preservative'. Advertising had helped transform antibiotics in the popular imagination. No longer were they simply cures; they were guarantors of health. Their many uses for prevention were referred to as antibiotic prophylaxis, from the ancient Greek word *phúlaxis*, the act of guarding. But who was guarding the guards?

It certainly wasn't the companies selling antibiotics. Pfizer's Terramycin was in competition with Lederle's almost-identical Aureomycin, each company touting its own patented molecule. And because tetracyclines were easy to find by screening soil, Lederle and Pfizer soon had other competitors. By the mid-1950s, 40 per cent of antibiotics sold in America were 'broad-spectrum' varieties that companies encouraged doctors to give as treatment for almost any bacterial infection. At the heart of the market were the tetracyclines. Their similarity was disguised by their various brand names – Terramycin and Aureomycin, but also Achromycin, Tetracyn and Polycycline. Though to patients and doctors they sounded like different medicines, they were medically interchangeable, with each available form producing resistance to the others as well. Even their supposed benefits had not been rigorously tested, because under America's permissive drug regulations, that wasn't necessary.

As more competitors joined the market, the patent situation became even more complicated. The various filings and documents on similar molecules created a legal hall of mirrors, a mess of subtle distinctions that was fertile ground where company lawyers could frolic. The various tetracyclines were all minor variations on a central molecule, tetracycline, which companies realised held distinct legal importance among the cat's cradle of manufacturing possibilities.

An almost incomprehensibly complex chain of events was set in motion: after realising that unadorned tetracycline gave a similar effect to Terramycin, Pfizer had filed a further patent claim seeking to protect their exclusivity. However, unadorned tetracycline had also been discovered in Texan soil by another company called Heyden, which had also filed a patent claim on this natural molecule. At this point, Pfizer's competitor Lederle bought Heyden and so acquired its claim to tetracycline. While the Patent Office was still considering the two claims and it was unclear which company would win the final patent, Lederle and Pfizer agreed a peace treaty. They held a private adjudication and agreed that Pfizer was the first inventor, Lederle withdrawing its claim. The outcome was that Lederle and Pfizer could both manufacture and sell tetracycline along with a third company, Bristol. Two other companies with their own brand-name versions could sell it, but couldn't manufacture it and would have to purchase it from the others.

Such a convoluted set of relationships between companies seemed impossible for government regulators to follow – and some believed that was precisely the point. According to one economist examining the agreements a decade later, the written agreement between Lederle and Pfizer was designed to permit behaviour which 'would have clearly been illegal' otherwise.

A patent was meant to be a prize to reward innovation. Yet it was plain to see that in the case of the tetracyclines, these new patents were not for inventions. Indeed, by identifying the same antibiotics, the companies involved had proved that anybody using the same methods would discover them too. Across America, companies were simply shaking the branches of the great tree of microbial life and collecting the golden lemons that fell.

The Patent Office's shift in stance to grant patents on products of nature conflicted with the traditional requirement that a patent show a 'flash of creative genius'. But changes were coming. Like the dye companies in nineteenth-century Germany, pharmaceutical companies were allowed to help with the drafting of a new patent law. The 1952 Patent Act enshrined the principle that 'patentability shall not be denied because of the way the invention was made'. The requirement for creative genius became a much weaker one of 'non-obviousness'. Companies could systematically find antibiotics using established techniques and be confident they could own them as intellectual property.

The golden lemons kept falling from the tree – but outsiders started to notice strange arrangements. According to the logic of capitalism, as more and more companies found similar antibiotics that differed subtly and all had different patents, they would compete like bacteria in the soil that shared the same niche. Simple economic theory dictated that the price of each antibiotic should start to fall. In practice, it didn't.

A sign that something was amiss was spotted by federal economist John Blair when he started comparing the prices of the new broad-spectrum antibiotics in 1953. Prices remained remarkably similar, even when other companies brought new competitor drugs onto the market. When Blair looked into the manufacturing processes for antibiotics, he uncovered the tangled mess of back-and-forth agreements that showed companies were collaborating rather than competing: the tetracycline cartel was the paradigmatic example. Between 1948 and 1952, the price of tetracyclines sold by Pfizer and Lederle had fallen by two-thirds as more companies entered the market, but the subsequent cross-licensing meant that prices stabilised over the following decade.

What had been presented to the world as a triumph of the free market was anything but. For all the talk of ruthless competition as the driver of capitalism, these arrangements were what ecologists would call symbiosis, a collaboration for mutual benefit. As Blair investigated, he found more and more evidence of the antibiotic cartel. The same companies who had participated in the US government's penicillin programme were now the leaders of the antibiotic craze. Their profit margins on the new broad-spectrum antibiotics were remarkable. The usual profit margin for a medicine was around 5 per cent. The new antibiotics routinely had profit margins of above 20 per cent – the highest for *any* type of drug – with one even having a margin of 52.1 per cent.

Part of the reason for this superb return on investment was that antibiotics were usually dirt-cheap to manufacture. After all, they were natural products made by bacteria, and the manufacturing methods had been perfected for penicillin and streptomycin with open collaboration, huge cross-industry investment and government support. But although the companies piggybacked on the ingenuity of bacteria to make cheap medicines, they weren't about to pass those savings on to the consumer. An antibiotic patent was a licence to print money.

Around the world, the boom in antibiotics caused problems for healthcare systems. When the National Health Service was founded in Britain in 1948, it had made drugs free on prescription. In the first year, expenditure on drugs was triple what had been expected, largely because of doctors prescribing far more penicillin than anticipated. As costs continued to spiral out of control, the Labour government eventually decided to start charging for prescriptions – a major violation of the initial aims of the NHS. The fallout caused the resignation of the minister Aneurin Bevan, the architect of the NHS, eventually leading to the collapse of the government.

Frustrated by the high price of the tetracyclines from American companies, both the British and the American governments tried to procure them from Italian manufacturers instead, who charged far less. The companies responded angrily. Pfizer claimed that the Italians had obtained the antibiotics by industrial espionage, stealing both bacteria and secret formulas, and sued the British Ministry of Health for infringing the tetracycline patent. Pfizer lost, but the message was clear: antibiotics were now serious business.

One might have thought that the vast increase in antibiotic prescription was improving people's health. But in 1959, Max Finland analysed infection rates at a typical hospital in Boston and found something alarming. Astonishingly, after 1947 and the introduction of broad-spectrum antibiotics, deaths were rising. There were now *more* people dying from serious bacterial infections. Somehow, antibiotics were driving new forms of disease. Gram-negative bacteria, including species that had only rarely been causes of serious infection, were now responsible for growing cases of fatal sepsis. Finland concluded that the destructive impact of broad-spectrum antibiotics on the 'total bacterial flora' (what we would now call the microbiome) had been severely underestimated. Antibiotics had transformed the treatment of certain infections, but their benefits were not universal, and they had hidden harms.

Companies continued to encourage doctors to prescribe antibiotics whenever they could. Finland had satirised the belief in endless prophylaxis, but the logic of broad-spectrum antibiotics as a cure-all was now extended even further: why not use multiple antibiotics together? After all, such combination therapy had revolutionised the treatment of tuberculosis. But although it could overcome resistance in a single pathogen, it was an unjustified leap to suggest that combining antibiotics was necessary or wise for otherwise healthy patients.

Combination therapy was attractive for another reason. Recombining antibiotics allowed for the appearance of a new product with a different brand name, creating yet another new medicine that could command a higher price. The logic of combination therapy was warped further beyond antibiotics alone: the other drugs could be anything, from vitamins to pain relief. If medicine was good for you, then surely *more* medicine was better.

By 1957, there were over sixty combined preparations of antibiotics available in the US. About half contained two antibiotics together – but some even contained five. That same year, a textbook on combination therapy warned that if the trend was not checked, 'rational chemotherapy will give way to chaos': a doctor would be 'confronted with such a bewildering array of antibiotic combinations supported by multicolored promotional material piling up daily'. The purpose of these new products was obvious to John Blair. As he wrote, these products were nothing more than attempts 'to capture again the high prices fading for the older products'. Yet every new combination was proclaimed by the company selling it as a major medical advance.

For its proponents, combination therapy represented a 'Third Era' of antibiotics. The phrase first appeared in a 1956 speech given by the man in charge of regulating antibiotics, the head of the FDA's Antibiotics Division, Henry Welch. Unknown to Welch's audience, the phrase had been inserted by a Pfizer copywriter because it was to be a central theme of Pfizer's subsequent advertising. Shortly after Welch's speech, he oversaw the approval of a new antibiotic from

Pfizer called Sigmamycin. Sigmamycin was simply a combination of tetracycline with another antibiotic, oleandomycin, but Pfizer presented it as the vanguard of the supposed Third Era: an article in the *LA Times* claimed that 'One+One=Three'. A Pfizer brochure promoting Sigmamycin claimed that 'every day, everywhere, more and more physicians find Sigmamycin the antibiotic therapy of choice.' Business cards from eight doctors were included, complete with addresses and phone numbers. But when a journalist attempted to contact them, he found something strange. The phone numbers were false; the addresses didn't exist. Neither did the doctors. A copywriter at Sackler's agency had simply made them up.

Welch was meant to be a regulator of the pharmaceutical industry, but he had become its willing collaborator: his speeches were secretly edited by Pfizer, and he was receiving hefty bribes. These payments were amateurishly disguised: after Welch had delivered his speech, Pfizer bought 238,000 copies from him. During the 1950s, Welch secretly earned more than $280,00 from the pharmaceutical industry (the equivalent of over $3 million today). In the absence of any effective regulation, the pharmaceutical sector had become the Wild West. But times were changing: in 1959, a new sheriff arrived in town, and he didn't like what he saw.

Senator Estes Kefauver was a powerful figure: big, raw-boned and slow-moving, peering at the world through his owl-like glasses. He was a Democrat from Tennessee who had taken on the Mafia as chair of a 1950 committee investigating organised crime, interviewing mob bosses live on television: from Louis 'Little New York' Campagna to Jake 'Greasy Thumb' Guzik and Paul 'The Waiter' Ricca. The hearings had made Kefauver a public hero. Now, he was turning his attention to the pharmaceutical industry. As chair of the Senate's Antitrust and Monopoly subcommittee, he wanted to find out whether pharmaceutical companies were deceiving the American consumer. Kefauver employed his long-time friend John Blair, the economist who suspected the companies of price fixing, on his committee staff.

As the hearings began, Kefauver grilled executives as if they were criminals. In Kefauver's view, they weren't benevolent life-givers – they were hucksters taking the public for a ride. The hearings highlighted questions they didn't want to answer. Why was a drug that cost $1.57 to produce and package being sold for $17.90? Why was the same product sold by different companies under different names, but at an identical price? And if companies justified their high profits by saying that was how they funded research into new drugs, why were they spending so much on advertising?

Pharmaceutical companies were alarmed by the sudden bombardment, referring to the first day of proceedings as their Pearl Harbor. Kefauver's committee also exposed the corruption of the FDA by publicising details of the payments Welch was receiving; Welch resigned the following day in disgrace. In its final report in June 1961, Kefauver's committee concluded that the fundamental structure of the pharmaceutical sector made it ripe for monopolies. It was dominated by a few large firms, with a patent system that ensured that companies could make unjustifiably high profits. Kefauver drew up a bill that would reform the sector, including sweeping changes to patents. The entire business model of companies like Pfizer was under threat.

The companies realised that to protect themselves, they needed to make Kefauver's proposed legislation seem not only misguided, but dangerous. To combat his plans, they formed a new institute whose sole purpose was propaganda for the drug industry. The Health News Institute promoted the 'drug story' which explained to the public how a drug was developed. Framed as a neutral summary, it emphasised the risk and expense of developing drugs alongside the benefits they brought, thanks to the American drug industry. The stress on 'American' was deliberate. Never mind that the two antibiotics used as illustrations – sulfa drugs and penicillin – had both been discovered in Europe, as Kefauver pointed out. In an age of McCarthyism and paranoia about Russian influence, linking the proposed reforms to socialism and communism made for a damaging accusation. Vice President Nixon warned that Kefauver's bill represented an attack on 'individual freedom'.

In the thirty-five pages of his bill, Kefauver set out a new vision for the pharmaceutical sector. The seventeen years of exclusivity a patent-holder currently enjoyed would be cut to only three. After that period, patent-holders would be forced by law to cross-license their patent for a fixed royalty fee of 8 per cent. In contrast to an encyclopaedia of brand names that concealed identical underlying products, only a single name would be permitted. And patents would no longer be awarded to a molecule simply for possessing a pedantically different structure – it would need to have a significantly improved therapeutic effect compared to existing medicines. In every respect, these rules had been informed by the story of tetracyclines.

Other politicians weren't as fired up as Kefauver about the issue. The bill had been kicked into the long grass in 1961, but Kefauver brought it back in March 1962. This time, the pharmaceutical industry was prepared. Kefauver had been assured by the Kennedy administration that the bill had the President's support. That wasn't true: in June, a special group assembled in a government conference room to rewrite it completely. Kefauver was not invited – but present and contributing were three representatives of the drug industry. The secret meeting gutted his bill, deleting all the patent provisions. Out of the thirty-five original pages, only fifty-five lines were left untouched. When the edited bill was revealed, Kefauver was outraged. It was a 'mere shadow' of what he'd intended.

One area where the drug companies were more willing to accept reform was safety. In the spirit of free enterprise, throughout the 1950s, new medicines like antibiotics had been approved and sold within months, well before any rigorous safety trials could have been conducted. All antibiotics tended to have some side-effects. Sometimes these were relatively benign, but others took time to become apparent. The tetracyclines turned out to permanently stain teeth if they were given to young children: they bound to calcium, staining the bright white American smile a dirty brown. Furthermore, antibiotics in development were often so impure that patients had allergic reactions to the impurities – but even pure penicillin provoked reactions in a minority. Some antibiotics had extremely rare side-effects that only showed up in a handful of people. Chloramphenicol, Parke-Davis's antibiotic discovered from Venezuelan compost, could be fatal in around 1 in 40,000 cases. That only became clear in the years following its licensing in 1949. Whereas now such a serious side-effect would lead to an instant withdrawal of the drug, chloramphenicol was still promoted and sold throughout the 1950s. Though doctors who had seen the side-effect often avoided its use, most hadn't and so continued to prescribe it.

In the 1950s, the FDA had only seven full-time drug reviewers. Faced with thousands of new medicines each year, they had little option but to wave most things through quickly. One new reviewer, Frances Kelsey, found that the safety information provided for her first assignment – an antinausea drug recently introduced in Europe – was insufficient. When she asked for more, she was given what she called 'an interesting collection of meaningless pseudoscientific jargon apparently intended to impress chemically unsophisticated readers'. She held firm, refusing to license the drug before more information was provided, specifically about its effects on pregnant women where it was recommended for reducing the symptoms of morning sickness. The drug was thalidomide. As the months passed, reports came in from Europe of terrible birth defects – thousands of babies had been killed or left with stunted limbs. It was only Kelsey's refusal to wave thalidomide through that had prevented the same happening in the US. Some children still died, because the company had given out over 2 million pills of thalidomide to more than 1,000 doctors for investigational use.

The public outrage when this was revealed in 1962 resulted in sudden safety reforms. Kefauver's edited bill was combined with another from a different senator into the Kefauver-Harris Act. Despite the act having his name on it, one historian writes that 'Kefauver's fingerprints were barely visible on the final bill.' The patent provisions were scrapped, but important changes gave the FDA increased powers to assess the efficacy and safety of drugs. One important positive consequence was that the FDA's reviewers realised that many of the antibiotic combinations on the market offered no benefit over individual drugs and revoked their approvals. But the new legislation also introduced more stringent requirements for clinical trials, which made drug development more expensive. That consolidated the ability to bring drugs to market in the hands of existing large pharmaceutical companies – which explained why they had been happy to go along with it.

During the 1950s, antibiotics helped pharmaceutical companies become one of the world's most powerful industrial forces. The ascent of Pfizer demonstrated how capitalism had harnessed microbes: from citric acid, to penicillin, through to Terramycin and beyond. Smith's deathbed advice to McKeen proved prescient. Like Pfizer's one-time opponents in the Italian lemon farms, they now controlled a crucial resource. Price-fixing lawsuits resulted in companies choosing to pay out millions of dollars in settlements without admitting liability. In 1968 the US government itself sued Pfizer and four other tetracycline companies. But after more than a decade in the courts, the government was ultimately unable to convince a judge that Pfizer representatives had a 'specific intent to defraud' when filing the tetracycline patent; the government lost. Yet by this point, to its most severe critics, the pharmaceutical industry had long since come to resemble a form of organised crime.

After an age of scarcity, the tetracyclines marked an age of excess. Antibiotics percolated everywhere, including back where they had come from: as the years passed, the abundance of tetracycline resistance genes in soil grew many times over thanks to environmental pollution. Broad-spectrum antibiotics were the flagship products of the developing American pharmaceutical industry; aggressive advertising set expectations high. Since many had few side-effects and were taken for mild illnesses – or even when patients were healthy – they became an accepted part of medicine. But there was no evidence that this proliferation was improving health. In fact, the widespread use was selecting for resistant bacteria that then circulated between people, creating truly serious infections that were harder to treat. One doctor wondered openly whether the excess in prescribing would come to be regretted: 'Are we in medicine, like our counterparts in industry, exhausting our most valuable resources at too rapid a rate?' But once doctors and patients had grown accustomed to antibiotics, it was difficult to imagine a self-imposed scarcity.

Advertising embedded the idea that it was negligent not to take antibiotics. Adverts promoted antibiotics for almost every condition including the common cold, an affliction caused by any one of several viruses which antibiotics are completely ineffective against. It was in 1950s America that cold remedies consisting of combinations of antibiotics first became popular, rising to more than 4 million new prescriptions in the early 1960s – despite no evidence at all that they were effective. The myth that antibiotics can treat a cold endures today.

During the tetracycline period, the pharmaceutical industry shaped how it was legislated. Companies thwarted attempts to curtail patent rights and restrict their profits, and were even invited into the room to rewrite government legislation. Other industries noticed these efforts and emulated them. William Kloepfer, the head of public relations at the Pharmaceutical Manufacturers of America (PMA), had overseen the fight against Kefauver's proposals, using techniques such as contesting scientific evidence, watering down new legislation and arguing that fundamental freedoms were at stake. Those tactics served as a playbook for other industries to learn from. In 1967, Kloepfer moved on to work for the propaganda wing of another industry, the Tobacco Institute. Tobacco companies and fossil fuel companies would come to use similar techniques to fight against their opponents in the public sphere.

Kefauver had done his best to rein in the pharmaceutical companies, but they defeated his efforts. He must have had a certain grudging respect for his opponents. After 'his' bill had been passed into law in 1962, Kefauver bought fifty-one shares in a corporate success story: Pfizer.

Thesis, antithesis
Amoxicillin and clavulanic acid

*Nature designed the penicillin molecule to teach organic
chemists a little humility.*

John Sheehan

The hospital is a paradoxical place: a building where people go to get better which sometimes kills them instead. In 1841, the Hungarian doctor Ignaz Semmelweis was working in the Vienna General Hospital when he noticed a strange discrepancy. Women giving birth in the clinic where medical students were taught died at a rate three times that of the neighbouring midwifery clinic. Semmelweis realised that the same medical students who assisted at autopsies were going on to deliver babies, their hands transferring what he called 'decaying animal organic matter' which somehow infected their patients. Semmelweis's work wasn't widely accepted at first, but over time doctors came to understand the need for antiseptic procedures. For the ever-present danger of a hospital was that by gathering sick people together to treat them more efficiently, doctors also created efficiencies for bacterial transmission. The medical staff who passed between patients could not only heal the afflicted but infect the uninfected. The widespread use of antibiotics in hospitals throughout the 1950s created a similar paradox. Medical care had become better than ever – but the hospital had now become a perfect breeding ground for antibiotic resistance.

This was particularly obvious for penicillin. The main concern was staphylococci, the bacteria against which Fleming had first observed the antibiotic's effects. Staphylococcal outbreaks in hospitals were becoming both more common and more resistant. In March 1956, two babies in Bristol in south-west England had been transferred to the city's Children's Hospital. When doctors sampled pus from the babies' chests, they found staphylococci that were resistant to penicillin, streptomycin and tetracycline. One of the babies died. By the end of April, four more were infected; by May, another ten. The doctors quickly realised their hospital had become the site of a resistant outbreak. The staff started trying to rid the hospital of the resistant strain, which could linger on surfaces ready to be acquired by a new person. Everything was to be cleaned: walls and floors scrubbed, sheets washed at high temperatures. But washing the heavy woollen blankets would ruin them. In the post-war austerity of 1950s Britain, throwing good blankets away was unconscionable, so the staff came up with an improvisation to kill the staphylococci: they loaded eighteen blankets at a time into a pressurised steam chamber and gassed them with formaldehyde. Their disinfecting procedures worked. June seemed to be the peak of the outbreak, and by October there were no new cases. Their experience was just one among many worldwide, as penicillin resistance in staphylococci continued to increase.

The bacteria could hide everywhere, from curtains to light fittings to even the pencils in a doctor's pocket. When a global pandemic of influenza virus struck in 1957 – the so-called 'Asian' flu – people who were seriously ill with the virus often contracted staphylococcal pneumonia, their weakened immune systems unable to prevent bacteria occupying their lungs. Many of these infections were undoubtedly acquired while patients were lying sick in hospital, the very place where they ought to have been most protected. In 1958, an American medical conference heard that up to 9 per cent of clean surgical wounds would be infected in hospital with staphylococci. Certain strains were now resistant to multiple antibiotics. Some doctors worried openly that the end of antibiotics was a possibility, with one British doctor putting it in military terms: 'We may run clean out of effective ammunition. Then how the bacteria and moulds will lord it.'

By 1960, penicillin resistance in London had become a major concern for doctors. The different responses to the resistance problem could almost be separated by psychological disposition. The pessimists sought to safeguard existing antibiotics: in one London hospital, penicillin was withdrawn from routine use in all but four wards. The optimists hoped for new antibiotics to replace them. In a London that was becoming increasingly cool and cosmopolitan, there were a lot of optimists. As one character told another in the novel *Absolute Beginners* (1959), 'This is London, not some hick city in the provinces! This is London, man, a capital, a great big city where every kind of race has lived ever since the Romans.' Celebrities from around the globe could be found in the British capital. They included Elizabeth Taylor, perhaps the most glamorous woman in the world, who had been in London since the autumn to film *Cleopatra*. It was the first time ever that an actor had received an upfront salary of $1 million.

Studio executives at Twentieth Century Fox had decided to film the ancient Egyptian epic at Pinewood Studios, eighteen miles west of London, encouraged by generous tax breaks from the British government in its efforts to compete with Hollywood. The film's director had been under the impression that filming would be in Rome. It was only when he flew into London and was driven to Pinewood for what he believed was a script meeting that he saw palm trees from the car window and realised he'd been double-crossed. Ancient Alexandria had been recreated in suburban Buckinghamshire at a cost of $600,000, stretching for over twenty acres. As well as the imported palm trees, there were opulent temples and fifty-foot-high sphinxes. With his first glimpse of the set, the director turned green and vomited in shock.

London might have been cool, but it was far from sunny; the reality of the British autumn soon set in. Even when the rain paused for long enough to allow filming, it was often so cold that the actors' breath could be seen rising like steam. Taylor had fallen sick on the second day and developed a fever, leaving the production without its star for weeks. After two months of filming had produced just ten and a half minutes of usable film, the director resigned in despair. The film's insurer, Lloyd's of London, expressed concern about the delays and suggested that the ailing Taylor should be replaced by Marilyn Monroe.

That concern would prove to be justified. On 4 March 1961, before filming was due to resume with a new director, a gasping Taylor turned blue and fell unconscious in her London hotel suite. She had been infected with penicillin-resistant staphylococci. On her arrival in hospital, it became clear that she had a severe case of staphylococcal infection in both lungs. The newspapers waited anxiously for updates, reassuring the public that her medical team included Queen Elizabeth's personal physician. To allow Taylor to breathe, the doctors performed an emergency tracheotomy, cutting a hole in her neck and using an electronic lung to supply her with oxygen. Doctors treated her resistant infection with a brand-new antibiotic called methicillin, a derivative of penicillin which had only become available the month that *Cleopatra* started filming. Five days after Taylor's arrival in hospital doctors announced that she was 'out of danger'.

Methicillin had saved Taylor's life, but the experience left her weak, exhausted and unable to talk. She was clearly incapable of acting, and Fox had no choice but to suspend the production, finally defeated by London's inclement conditions – both its weather and its penicillin-resistant staphylococci. They dismantled the $600,000 set to ship back to California and began a long conversation with Lloyd's about claiming on their insurance.

On 28 March, Taylor travelled to the US, descending the ramp of her plane in a wheelchair, and less than three weeks later, she attended the 33rd Academy Awards in Hollywood. She wore a pale-yellow bodice and an ivory bubble skirt with green flowers of embroidered silk tumbling down it, a green belt wrapped around her waist with a pink rose tucked in. When it was announced that she had won Best Actress (for the film she'd completed before *Cleopatra*), she covered her face with her hands then walked slowly to the stage. Only then was it apparent just how frail she was. She whispered her brief acceptance speech, pausing to take deep gasping breaths. On her neck, just visible beneath her makeup, was the thin vertical scar where the surgeons had cut open her neck. Production of *Cleopatra* resumed in Rome later that year, with Taylor still playing the title role. Without methicillin, she would probably have died.

The antibiotic that had saved Taylor's life marked a radical moment in the history of antibiotics. Until then, there had been a divide between synthetic antibiotics, such as Prontosil or isoniazid, and natural antibiotics produced by bacteria or fungi, such as penicillin. But the discovery of methicillin built a bridge between the two. Beginning with the structure of penicillin as their blueprint, chemists had designed a brand-new antibiotic, as powerful as natural penicillin but evading the problems of resistance. The process they had followed could, they hoped, deliver unlimited new ammunition against resistant bacteria.

John Sheehan was an American chemist, a brash and confident man with a high opinion of his own abilities. As a chemist at Merck in the 1940s, when streptomycin was in its earliest stages of production, he'd asked if some could be spared to treat his mother who had a difficult bacterial infection. He was told that Merck was struggling to purify the drug and patients were having allergic reactions to its impurities. Looking over the methods his colleagues had tried, he recalled an old German chemical process for purifying sugar. Trying it overnight on streptomycin, he was told he'd just purified the highest-quality version yet of the antibiotic. He noted with amusement that the problem had taken him only a day to solve. Penicillin proved a far greater challenge. Like many other chemists, Sheehan had presumed that the best way to make huge quantities of the drug would be to make it chemically using established procedures from raw ingredients, not by brewing microbes. But despite enormous wartime efforts involving over a thousand chemists, they had failed. By the end of WWII, fermentation vats were making enough penicillin the natural way, and synthesis efforts were largely abandoned. Whatever the *Penicillium* mould was doing inside its cells, it was far beyond the capabilities of industrial chemistry.

Chemists had proposed over ninety different structures for penicillin before the correct structure was obtained by Dorothy Crowfoot Hodgkin at the University of Oxford in early 1945. Hodgkin had started working on the challenge in a 'disgustingly cheerful' mood, but it had proven to be painstaking and difficult work, firing X-rays at delicate crystals of penicillin and poring over the results. The tentative structure she eventually inferred stretched credulity. The atoms were arranged in a scarcely believable ring of molecules, apparently bound together under such strain that the result violated many chemists' intuitions. To confirm her structure, Hodgkin negotiated access to a gigantic punch-card machine – an early computer – in order to run the intensive calculations required. The machine was normally used by the government to keep track of ships' cargoes. Her co-workers wrote a set of computer commands for the calculations and sent it to the machine's operators to run, but when the first results came back, they were garbled. It seemed that the operators, uninterested in the scientific research, had been sloppy and careless when they inputted the commands. After the scientists submitted identical commands, but this time with a title that disguised them as relating to a fictional ship, the machine's operators ran them properly. The results were unequivocal: the ring structure was correct.

Penicillin's structure was solved, but not everyone was happy about it. Hodgkin had made a final model of penicillin from wire and cork that revealed the strange ring at its heart. When she was taking it to show another Oxford chemist, Ernst Chain warned her, 'You'd better take a gun with you'. The molecular arrangement was so unexpected that one chemist had reportedly said if it turned out to be real he'd give up chemistry and grow mushrooms instead. But it was real, and it became known as the beta-lactam ring. The ring was, in Sheehan's words, 'a diabolical concatenation' beyond the toolbox of human chemists. It was now clear why everyone's efforts to synthesise penicillin had failed; the beta-lactam ring was like an artefact from a more technologically advanced civilisation. Sheehan compared synthesising penicillin to trying to repair an expensive watch with 'anvil, hammer and tongs'. But Sheehan believed that chemistry would one day be capable of the seemingly impossible task. When he moved from Merck to MIT in 1948 to become an academic scientist, he decided to use his newfound freedom to take it on himself.

Sheehan enjoyed working on a difficult problem where he had few competitors. When else, he thought, would he have such green pastures entirely to himself? He signed an agreement with the American pharmaceutical firm Bristol, similar to Waksman's deal with Merck: Sheehan would act as a consultant, and in return, Bristol would finance his research. Penicillin synthesis seemed like a pointless task when production was booming by natural fermentation, but Sheehan reasoned that if he could make penicillin, he could also make variants of penicillin. Why should natural penicillin, evolved in soil for fungi, be the apotheosis of antibiotics? Penicillin was a miracle, but it was far from a perfect drug. If chemistry could master its structure, then nature's bounty could be improved: a chemist could improve penicillin's absorption in the body, enable it to survive stomach acid so that it could be swallowed rather than injected, or perhaps even make it more powerful. As reports of penicillin-resistant staphylococci became more frequent, Sheehan also speculated that by surrounding the beta-lactam heart of penicillin with a shield of other atoms, he could protect it from penicillinase. Penicillinase is the name given to any enzyme produced by certain bacteria, including staphylococci, that can break apart the penicillin molecule by striking at its beta-lactam ring. Such enzymes were a key mechanism in penicillin resistance. By modifying antibiotics to defend against penicillinase, Sheehan believed chemists could solve the resistance problem.

After almost a decade, by 1956 he had figured out how to synthesise the beta-lactam ring of penicillin on its own, a chemical substructure that was given the shorthand 6-APA. This was a huge achievement. The next stage was to work out how to build out from that substructure to get all the way to penicillin itself. In

1957, he announced that he'd succeeded. In essence, he first made 6-APA and then ran reactions that delicately stapled on the other parts of the molecule without disturbing its unstable bonds. He had his proof of principle. The process was incredibly expensive compared to fermentation, but it was possible. Sheehan could replicate synthetically what Fleming's *Penicillium* had evolved to do naturally. But by now he was no longer alone on his green pastures. A British company had stumbled by accident onto the potential of producing new forms of penicillin.

Beechams was an unlikely entrant to the field of antibiotic development; its products included laxative pills, toothpaste, Brylcreem hair gel and the energy drink Lucozade. But its managing director had realised that the founding of the National Health Service in 1948 had made prescription medicines a growth market, and none more so than antibiotics. He wanted to remodel the company along 'American lines', following the example of the pharmaceutical firms that aggressively advertised their products to doctors directly.

Beechams' research division was based at Brockham Park in the countryside outside London, a jumble of buildings around a repurposed stately home, similar to Pinewood Studios, where Fox had recreated ancient Alexandria. In 1954, Beechams had enlisted Chain, the chemist from Oxford who'd first isolated penicillin, as a consultant on antibiotics. With Chain's encouragement, the chemists at Beechams started investigating penicillin, building their own fermentation plant where they recorded a puzzling anomaly. During fermentation by *Penicillium*, there were two ways of measuring the amount of penicillin in the broth. You could measure the antibiotic activity against bacteria or you could test chemically for the presence of the beta-lactam ring alone. These methods should have been different ways of getting the same result, but consistently the chemical test gave higher results. Beechams scientists took the anomaly seriously: it told them there were more unfinished beta-lactam rings in the broth than there were complete penicillin molecules. Just like in Sheehan's process for synthesising penicillin, the *Penicillium* were first making 6-APA then adding the finishing molecular touches. Within six months, Beechams had managed to extract crystals containing 6-APA. All those years that Sheehan had been working out how to make 6-APA chemically, the penicillin industry had been pouring millions of dollars of it down the drain.

If Sheehan was unnerved by the sudden presence of a new competitor, he didn't show it. He claimed he'd synthesised 6-APA before Beechams. Bristol, the company funding his research, made an agreement with Beechams in April 1959. The two companies would collaborate on making new varieties of penicillin, both by modifying the naturally produced 6-APA from fungi and by creating new molecules directly from raw materials.

Penicillin had been singular, but it was about to become plural: the penicillins were a family of molecules, all sharing the beta-lactam ring, just as the tetracyclines had shared their own central structure. Beechams' first semisynthetic penicillin, a molecule called phenethicillin, was launched in October 1959 with a press briefing held, to the outrage of the British medical establishment, not at a medical conference but at a fashionable London restaurant. Beechams trumpeted the claim that it was twice as effective as penicillin, but their promotional leaflet was thin on published evidence – the BMJ complained that the so-called bibliography consisted only of items listed as 'personal communication'. The American method of pharmaceutical advertising was taking hold.

Making new penicillins was now straightforward. At the dedication ceremony for a new antibiotic factory, Bristol got the Governor of New York to turn a stopcock and personally synthesise a brand-new penicillin. Bristol had double-checked with patent lawyers beforehand that the Governor would have no legitimate patent claim if the antibiotic turned out to be useful and he tried to sue for his own status as an inventor. 'He is merely unskilled labor', the lawyers reassured them.

Beechams now developed a relative of penicillin that was immune to penicillinase, creating an altered molecule that protected the beta-lactam ring. They called it methicillin. For methicillin's first safety test in humans, two of Beechams' research directors volunteered. As if in a duel, a doctor and a microbiologist stood facing each other armed with syringes. First the doctor injected the microbiologist and waited to see if he needed urgent medical treatment. (They had agreed this order made the most sense.) When nothing seemed to have happened after half an hour, the microbiologist injected the doctor. They were satisfied enough to give the drug to other people.

Methicillin was on sale just eighteen months after Beechams' chemists had first made it. Two doctors at one London hospital reported cheerfully that they had sprayed an entire ward with the new antibiotic to try to eliminate all the staphylococci. They recommended others do the same. The risk of methicillin resistance was, they wrote, a purely 'theoretical objection'. Though they admitted they couldn't prove that methicillin resistance was impossible, they assured their colleagues it seemed highly unlikely.

Methicillin had saved Elizabeth Taylor's life, but four days before *Cleopatra* had started filming, Mary Barber had published a short letter in a medical journal cautioning that it was unwise to conclude that resistance was impossible. Barber had found she could make methicillin-resistant staphylococci herself in the laboratory with exposure to the antibiotic. She noted that it had also been said that resistance to tetracycline wouldn't be a problem; bacteria had disproved that claim. She advised doctors to use methicillin sparingly. In response, George Stewart, a doctor at a children's hospital, wrote in arguing that her results had no medical significance for the time being. After all, her research was based only on evolution in the laboratory, not the hospital. 'The proof of the pudding is in the eating,' wrote Stewart, 'and it seems premature to dwell on the hypothetical effects that may occur in the future as a result of over-indulgence rather than on the undoubted fact that the pudding itself seems to be a very good one for which we should be truly thankful.'

But within a few months, Barber had found naturally occurring staphylococci that could grow in the presence of methicillin. Methicillin was not invincible. In May 1961, Stewart – the doctor who had downplayed her results – found that one of his patients was infected with a methicillin-resistant bacterium. The outbreak spread to seventy-five children in the hospital, and one child eventually died of blood poisoning. The resistant strains had started at such low levels that they wouldn't have been noticed if he hadn't been looking. He urged doctors to be vigilant and get infected patients out of their hospitals. Barber's fears had been justified, and Stewart admitted he had to 'eat some if not all of his own words'.

That same year, scientists including Chain and Sheehan gathered at an industry-sponsored conference on resistance to penicillins. Twenty years before, the original penicillin had been the world's best medicine, standing alone. The news that the beta-lactam ring had been mastered and used to spawn new antibiotics was rippling through the pharmaceutical industry. If soil-screening had been the drug discovery machine of the 1950s, at the start of the 1960s it looked as if the baton would be handed back to chemistry, with the promise of outdoing nature itself. Some attendees were in the pessimistic camp – surely these new antibiotics would be overcome by resistance too? But Sheehan fought back. He didn't deny that chemists were in an arms race against the apparently never-ending adaptability of bacteria. But, he argued, chemists were up to the challenge. The more chemists learned from antibiotics, the more they could use that knowledge against bacteria. It wasn't just penicillin that could change.

In the city of Cagliari on the island of Sardinia, the streets run down to the Mediterranean from the old castle that hulks above the bay, and halfway down the Via Porcell is the Cagliari School of Hygiene. Its director, Giuseppe Brotzu, was a short and sombre man in his fifties who dressed in dark suits even at the height of summer. Locals called him *l'uccello del malaugurio*, the 'bird of ill-omen'. But Brotzu had quietly done his best to bring good fortune to Cagliari, tracing the spread of *Salmonella typhi*, the bacterial species that caused typhoid fever, through the city's jumble of sewers to understand its routes of transmission. Dismayed by the way that sewage poured straight into the sea, he had discouraged people from eating local mussels, but was surprised by how few people swimming in the harbour seemed to get sick. Inspired by the news he had heard of antibiotic-producing moulds like *Penicillium*, he tried to discover whether Cagliari's sewage contained organisms that could kill pathogenic bacteria. After experimentation, he isolated a mould capable of killing many different bacteria, including those that caused plague, dysentery and cholera. He injected his mystery mould juice into patients – without any animal tests beforehand. Though he observed side-effects like severe fevers, probably caused by immune reactions to impurities in the juice, he thought his results were promising, and hoped that his *antibiotico* could one day be extracted.

Post-war Cagliari was not the place for antibiotic development. In 1948, Brotzu sent a copy of his results to a British medical officer he'd met during WWII who had been stationed on Sardinia. The officer, once back in England, passed it on to Howard Florey at Oxford. Research into Brotzu's mould was slow, distracted by ongoing work on penicillin, but by the mid-1950s the Oxford scientists had confirmed that the Cagliari fungi could produce several antibiotics. They called these molecules cephalosporins after the *Cephalosporium* that produced them, just as *Penicillium* had given them penicillin. In 1959, they found that one particular cephalosporin contained a beta-lactam ring similar to penicillin, but was nevertheless immune to penicillinase. The cephalosporin therefore presented a tactical alternative: though it was far less powerful than penicillin, it offered another route of attacking penicillin-resistant bacteria.

The Oxford scientists formed a 'Cephalosporin Club' with a group of manufacturers. The experience with penicillin proved salutary: rather than working with the original mould Brotzu had sent, thousands of mutant strains were screened to find those that made more of the antibiotic. The problem was that modifying cephalosporins was harder than modifying penicillins, because the cephalosporin version of penicillin's beta-lactam heart proved much harder to make. Chemists at Eli Lilly's Indiana research laboratories solved the problem in 1962, allowing the cephalosporins to be pursued as a source of new antibiotics too. From the mid-1960s, American firms began to bring new cephalosporins onto the market.

The new synthetic penicillins and cephalosporins offered huge improvements. Not only could they be made immune to penicillinase, but some of them had increased killing power. Unlike the newer broad-spectrum antibiotics such as the tetracyclines, penicillin had only been effective against Gram-positive bacteria. But in July 1961, Beechams published details of its latest antibiotic, ampicillin. Made by modifying the natural beta-lactam ring, it was effective against a wide range of bacteria – across even the fundamental division of bacteria that Hans Gram had discovered with his stain in the 1880s. Ampicillin killed the Gram-positives like staphylococci, but also the Gram-negatives like *E. coli* that were causing rising numbers of deaths. Beechams, the second-largest advertiser in Britain, now had a pharmaceutical product that was an advertiser's dream, combining two miracles in one: a broad-spectrum penicillin. They set ampicillin's initial price at twice that of tetracycline. By 1970, Beechams' pharmaceutical sales had increased by fifteen times, and plants around the world were manufacturing their new penicillins.

In 1972, the science fiction author Isaac Asimov drew on these developments to strike an optimistic view in the face of the problems of resistance. Like Sheehan, he drew a comparison between human and bacterial adaptation: bacteria certainly had a 'stubborn versatility', but so did chemists. Asimov was confident that the chemists could keep humanity ahead in the evolutionary arms race. A new antibiotic would always be followed by resistance, but the chemists would always create a new antibiotic to replace it, and so on. 'Presumably,' he wrote, 'the merry-go-round will go on forever.'

Asimov was right that bacteria were stubbornly versatile. Almost as quickly as scientists could introduce new antibiotics, bacteria seemed able to summon new genes. In 1965, scientists had reported the discovery of a new penicillinase that was capable of breaking down the newly modified antibiotics. It had first been detected in Athens in 1963, carried by *E. coli* from a hospital patient. The new penicillinase came to be known as TEM, a shorthand for the strain it had been carried by. Beechams had recently created a new penicillin, amoxicillin, which could be taken by mouth rather than injected. It was an instant success with doctors – but not with bacteria. It too could be destroyed by TEM. TEM also conferred resistance to some of the latest cephalosporins. Soon, TEM had been seen in other bacterial species beyond *E. coli*. Ominously, the resistance gene was somehow able to cross species boundaries.

In February 1976, the first reports came from the port city of Liverpool, England, that TEM had appeared in patients with the sexually transmitted infection gonorrhoea, carried by *Neisseria gonorrhoeae*. Sailors came to Liverpool from all over the world, where doctors were accustomed to treating their sexually transmitted diseases. Just one injection of penicillin had always been a near-certain cure, but now it failed in every single patient. The TEM-carrying strain only made up 9 per cent of total gonorrhoea cases in Liverpool, but the doctors warned the proportion of untreatable infections would only rise as the resistant bacteria outcompeted the susceptible ones. (Indeed, by 1981, it would be spreading throughout Britain.)

Pharmaceutical companies like Beechams were concerned. Other enzymes beyond TEM emerged that could destroy many of the shiny new penicillins and cephalosporins. These powerful new medicines were wilting like lettuces in the midday sun. Rather than trying to develop another variant of penicillin that the enzymes couldn't destroy, Beechams' scientists began to investigate a new possibility: they would take the enzyme on directly. They were looking for a new sort of drug, an inhibitor that would block TEM's activity altogether.

In May 1977, Beechams announced that their chemists had found one. And yet again, it was a discovery from bacteria. The inhibitor was a molecule made by a *Streptomyces* species which had first been isolated from soil collected in South America by scientists at Eli Lilly. The Lilly researchers had called the bacterium *Streptomyces clavuligerus* because of its appearance: under the microscope they could see short side-branches of spores forming as if it was bearing little clubs ('clavulae'). The new molecule, which the Beechams scientists called clavulanic acid, bound tightly to TEM. Clavulanic acid was itself a molecular relative of penicillin and contained the beta-lactam ring, which was key to how it could bind to penicillinase. Beechams now proposed to combine clavulanic acid with amoxicillin into a single tablet that could be swallowed, dubbing the combination Augmentin.

Beechams' new medicine was a spectacular amalgam. Fungi had made a natural molecule, penicillin, which humans had learned to modify to make other molecules. As humans had used these antibiotics in greater quantities, more bacteria began to carry an enzyme that could destroy them. Then, by studying bacteria, humans had identified a natural molecule that blocked the enzyme. It was a Frankenstein creation bringing together the expertise of three kingdoms of life: human, fungal, bacterial. Augmentin was the first antibiotic that had been constructed with deliberate evolutionary logic, anticipating the bacteria's response like a judo master. It rebalanced the equation in humanity's favour.

Scientists had taken beta-lactam antibiotics out of the soil, concentrated them and put them into human bodies. Suddenly, bacteria for which soil had been a distant evolutionary memory had been exposed to massive amounts of beta-lactams. The result was a huge selective pressure in favour of those that evolved resistance. But they didn't need to reinvent the wheel – they could gain resistance by gaining genes that came, like the antibiotics, from soil bacteria. The emergence of penicillinases like TEM in clinical bacteria was a partial glimpse of a wider truth: bacterial species could swap genes. Research into antibiotic resistance would revolutionise the understanding of bacterial evolution, and make clear where their stubborn versatility came from.

The first inkling of the problem had come from Japanese researchers. WWII had left large parts of Japan in ruins, with thousands of people lacking adequate sanitation in bombed-out cities. Such awful conditions were ideal for the spread of bacterial diseases like dysentery, caused by a bacterium known as *Shigella*. After the Japanese surrender, the Americans had brought sulfa drugs to the country. These drugs worked at first, but in subsequent years, cases surged back up: sulfa-resistance in *Shigella* became common because of widespread use. Throughout the 1950s, further new antibiotics arrived in Japan: streptomycin, then chloramphenicol, and finally the tetracyclines. Dysentery cases dropped – but once again, they increased in subsequent years. So far, this made perfect sense: the more an antibiotic was used, the more resistance to it rose. But as long as multiple antibiotics were available, a successive handing on of the baton would still let doctors treat the disease.

At that time, it seemed that antibiotic research formed a virtuous cycle. Selman Waksman had first travelled to Japan in 1952, when he was still making enormous amounts of money from streptomycin. He had set up a trust to manage the proceeds, and arranged for the royalties from Japanese sales to support Japanese microbiology. The Waksman Foundation would award research funds from his patent share – the profits from each previous antibiotic would fund the discovery of the next. One Japanese microbiologist who received money was Tsutomu Watanabe, who was working on a better understanding of antibiotic resistance in dysentery. Before the end of the 1950s, he had made a startling discovery that upended assumptions about the very nature of antibiotic resistance.

Resistance was meant to be a response to antibiotic treatment. If a patient with dysentery carried *Shigella* that showed no resistance, and you then gave them a single antibiotic – say, chloramphenicol – you'd expect to then be able to isolate chloramphenicol-resistant *Shigella* from their body. But when Japanese microbiologists looked at such cases, they found something bizarre. The bacteria they isolated were *also* resistant to sulfa drugs, streptomycin and tetracycline. These were four completely different antibiotics, but resistance to all four had emerged after using only one of them. In a controlled laboratory environment, isolated bacterial species developed resistance only to the single drug they were exposed to;

gaining multiple resistance required multiple exposure. So why was it different in living patients?

Through careful work, Watanabe and his colleagues were able to explain this puzzling observation. Rather than thinking of a patient as carrying only one bacterial species, it was important to remember that their gut contained hundreds of different bacteria – not just *Shigella*, but many others, such as the closely-related *E. coli*. Isolating a single species from a faecal sample was like shining a narrow torch beam into a darkened room without moving it around. Watanabe broadened the beam, isolating other species from the samples. He didn't have far to look.

Where multiple resistance could be seen in *Shigella* after the patient had taken chloramphenicol, Watanabe could find it in samples *before* the antibiotic treatment in *E. coli*. Somehow, the resistance had moved between the two bacteria. Four different forms of resistance required four different genes, but they were somehow encapsulated in a transferable unit of resistance that could flit between species. That transferable unit of resistance must have arisen from the previous use of antibiotics in different patients, but it had been able to remain lurking within the gut microbiome. There were many species that could hide it. And when a patient took chloramphenicol, any *Shigella* in their gut that by chance picked up the transferable resistance through exchange would survive; the *Shigella* would then also pick up those three other resistance genes for free. Watanabe's transferable unit could instantly turn a sensitive bacterium into one able to resist all available treatment. The human gut was a melting pot where bacteria exchanged ideas in the dark.

The work had initially been published in Japanese journals, but after being publicised by the British microbiologist Naomi Datta, it was published in English in a series of papers in 1961. Up until this point, researchers had assumed that resistance to antibiotics could be sequentially addressed. When resistance emerged to one antibiotic, you could switch to another: from streptomycin to chloramphenicol to tetracycline. With enough antibiotics in the arsenal, science would simply outgun the bacteria – and with combination therapy, you could make extra sure that it was almost impossible for resistance to emerge. But *multiple* drug resistance that was encoded on a single transferable unit torpedoed that idea. Bacteria could casually acquire a crowd-sourced memory of all the antibiotics that had been previously used – not even necessarily against their own species. Doctors had presumed that they'd be starting with a full magazine of bullets against every new infection. Watanabe's discovery meant their magazine might be full of blanks.

The appeal of the broad-spectrum antibiotics was that they were indiscriminate – they worked against many different bacteria. But the confirmation of transferable resistance sharpened a huge evolutionary sting in the tail of that success. Unlike a narrow-spectrum antibiotic such as penicillin, a broad-spectrum antibiotic selected for resistance not only in the bacteria causing the infection – if there even was one – but in many others. Whether those bacteria were in the human gut, or in wastewater, or in the environment, dosing them with broad-spectrum antibiotics meant that more resistant variants were likely to thrive.

Watanabe had demonstrated that multiple forms of antibiotic resistance could be transferred all at once between different bacteria. As terminology wobbled and settled, the favoured term for these units of transfer would become 'plasmids'. Most bacterial species have the majority of their genome concentrated in one large loop of DNA: the chromosome. If the cell is a computer, the chromosome is akin to its operating system. In that metaphor, a plasmid would be an app: a much smaller and more agile piece of software. Plasmids can't run the cell, but they can let it do something *new* without changing the operating system. They are particularly suited to carrying genes involved in quick adaptation to an environment. Bacterial genomes are not static, but constantly in flux. Lines of code that are in an app can be copied and pasted into the operating system, and vice versa. But, all else being equal, a gene that is fundamental to a bacterial species should, over long evolutionary time, find its rightful place in the chromosome.

Plasmids, in contrast, are where evolutionary experimentation is at its wildest. Whereas there is typically only one chromosome in an average cell, a plasmid can be present many times over. That means plasmids increase the rate of resistance in two ways: by providing many copies of the same gene, and by allowing the gene to mutate at a higher rate, finding beneficial mutations more quickly. Plasmids were an ancient evolutionary innovation that had always allowed bacteria to rapidly adapt to changing pressures; the introduction of antibiotics was no different. The existence of plasmids made the notion of species more complicated in bacteria than in animals. Biologists had traditionally defined species by reproductive isolation: if two specimens could not mate and produce fertile offspring, they could not mix their genes and therefore were not of the same species. But bacteria separated by huge evolutionary distances – more genetically different than humans and fish – can still sometimes exchange genes with each other. It is as if, while drowning, you could rub up against a salmon and acquire gills.

Plasmids meant that bacteria could acquire resistance not gradually, as Fleming had claimed they could be 'educated' to do, but all at once, even to multiple antibiotics at the same time. For every antibiotic that was made naturally by microorganisms there would be a gene somewhere that could confer resistance. Though bacteria that were adapted to humans might lack that gene, it could conceivably make its way to them via a plasmid. If a patient was prescribed an antibiotic for an infection caused by species A, the memory of antibiotic exposure could be lurking somewhere nearby in their microbiome on a plasmid in species B. Taking the antibiotic would lead to populations of species A carrying the plasmid too.

In theory, once a patient came off antibiotics, that should have led to the rapid disappearance of resistance genes, since without antibiotics the genes should serve no benefit, and indeed might harm the bacteria's growth. But for the most part, the resistance genes didn't disappear. There were any number of reasons. Perhaps it was because the parasitic nature of plasmids meant that they were adept at hanging on within the bacterial cell, or at moving into other species, ready to jump back aboard from this reservoir. Perhaps once the bacteria acquired the resistance genes, they evolved to accommodate them. Or perhaps antibiotic overuse and environmental pollution meant that antibiotics were often present at low levels, making the genes more beneficial in the real world than they appeared in a test tube. Whatever the combination of reasons, the observations were indisputable: as resistance genes spread through bacterial populations, they tended to stick. Biologists watched in horror and admiration as plasmids snowballed in size, combining more and more resistance genes together: toolboxes for survival in the antibiotic era.

Bacterial genomes are layered like palimpsests, medieval manuscripts where one document has been written over another older one: in the Middle Ages, manuscripts were often imperfectly cleaned first, leaving traces of the older handwriting still visible beneath the most recent document, allowing historians to reconstruct the document's history through close reading. In the same way, by reading bacterial genomes, scientists found they could reconstruct the history of twentieth-century medicine: first came the genes to resist the disinfectants used in hospitals from the 1930s, then resistance to the sulfa drugs, and then further resistance genes tracking the introduction of new antibiotics. Just as antibiotics had emerged from the dark corners of biology and chemistry to become an indispensable part of medicine, so too these obscure resistance genes had become part of the genomes of many different bacterial species. Hospital wards were environments rich in antibiotics, and with lax cleaning practices encouraged by an overreliance on antibiotics, they became hotspots for these multi-resistant bacteria. Once there, they could not easily be unstuck.

When chemists first beheld the beta-lactam ring, they had felt mystified. But by learning from its apparently impossible chemistry, they had built up an uneasy working relationship with bacteria. The idea of vanquishing bacteria with a single antibiotic was replaced with an understanding that this was an eternal exchange. While there had once been a strict division between the synthetic chemistry that had produced Prontosil and the natural magic of penicillin, the boundaries between the two disciplines became porous: antibiotic development could now proceed as a joint effort between human chemistry and microbial evolution. As the twentieth century passed, more varieties of natural antibiotics containing a beta-lactam ring were discovered in soil bacteria and introduced into medicine, from the monobactams in the 1970s to the carbapenems in the 1980s. These and other new beta-lactams could be modified further by chemists to stave off resistance or change their chemical properties, just as Sheehan had envisaged. It was a little like molecular jazz: bacteria had been playing the same tunes for millions of years, and chemists were simply improvising around these old standards. But bacterial evolution soon responded to these improvisations, and the counterpoint of antibiotic resistance could often be heard in the background: at first faint, but getting louder.

If the beta-lactam ring had once presented chemists with a seductive mystery, understanding it gave them a new affliction: hubris. Chemists could generate endless variations on nature's themes, creating a near-infinite number of 'new' antibiotics. Yet the apparent abundance of chemistry was illusory. Beta-lactam antibiotics became the most prescribed antibiotics in hospitals, but not a single one of the successive generations of artificial beta-lactams had the transformative impact of their molecular grandfather, penicillin. Asimov's antibiotic merry-go-round relied on an assumption of equality between humanity's actions and bacteria's response – an assumption that was violated by this slowdown of innovation. 'The ingenuity of nature', one microbiologist wrote about the beta-lactams, 'has not been matched by human beings.' The tools of chemistry might have become adept at altering antibiotics discovered from nature, but they could not conjure up wholly new ways of attacking bacteria.

'When a victory is won,' wrote the Chinese general Sun Tzu in *The Art of War*, 'one's tactics are not repeated.' Yet by relying so extensively on modifying extant antibiotics, that was exactly what the pharmaceutical industry did.

Last resort

Colistin

American pigs and Soviet pigs can co-exist.
Why then cannot our nations co-exist just as well?

Nikita Khrushchev

The end of the world has a long history. From the ancient Zoro-astrian doctrine of a final heroic battle between good and evil, to the thousands of devotees of the American preacher William Miller who waited in vain for the Earth to be purified by cleansing fire on 21 March 1844, to the Japanese doomsday cult Aum Shinrikyo who prophesied a cataclysmic nuclear war in 2003 – all societies have their share of people gazing towards apocalypse. In the twentieth century, antibiotic resistance became a potential new horseman. In 1963, Selman Waksman dismissed those warning about the end of antibiotics as 'gloomy prophets'.

It was a legitimate point. For all the dire predictions – from Fleming's warning in 1945 onwards – the long-prophesied end of the antibiotic era had repeatedly failed to arrive. As critics of Mary Barber's work on new forms of resistant staphylococci had pointed out in the 1960s, one observation of resistance to an antibiotic did not mean its immediate demise. The first observation of resistance would be an isolated case; even when resistance was seen in multiple patients, the antibiotic could still be relied on most of the time. But as resistance levels rose in a given hospital or area, doctors would begin to avoid prescribing it. When the problem reached epidemic proportions, as for penicillin-resistant staphylococci in the 1950s, the

157

antibiotic would be finally outmoded. But those concerned about the rise of antibiotic resistance were not prophesying a world where every bacterium could resist every antibiotic; that was a straw man. Antibiotic resistance was not a single conflagration but hundreds of slow-burning fires.

One of those fires was methicillin-resistant *Staphylococcus aureus* (MRSA). The first case had been recorded in London in 1962. It was a harbinger: over the following decades, MRSA became one of the most alarming public health problems in high-income countries, with strains gaining multiple forms of antibiotic resistance. From 1992 to 2002, the number of death certificates in England mentioning MRSA increased fifteenfold. At the start of the 2000s, separate epidemics in hospitals and communities in the US appeared to be converging, with only a few antibiotics left to treat it. MRSA could be carried by healthy people on their skin, but in rare cases, it could invade deeper into the body's soft tissues, causing virulent abscesses that could lead to fatal sepsis. Surveys showed that 1.5 per cent of Americans were carrying MRSA; in the nation's jails, it was up to 17 per cent. Witnessing MRSA's rise, one doctor who had worked in San Francisco in the 1980s judged that 'other than AIDS, this is the biggest thing in the past thirty years.'

By 2005, an estimated 278,000 people in the US were being hospitalised each year with MRSA. Over 6,000 died. That was a large number, but MRSA was just one among many possible bacterial infections. When all hospital-acquired infections were combined, they became the nation's sixth-leading cause of death. Tackling MRSA became a priority – not through new antibiotics but through better control: handwashing, sterilisation and cleaning. The 1950s doctors in Bristol who fumigated their hospital's woollen blankets to deal with a penicillin-resistant outbreak would have recognised many of these techniques.

The pharmaceutical industry was producing ever-more drugs each year, but it had failed to keep pace with bacteria. This could be seen by looking at the rate of introduction of new antibiotic classes – a class being a group of related antibiotics that work by the same mechanism. The number of antibiotic classes had grown rapidly between 1940 and 1970, but the rate of increase slowed dramatically in the 1980s. The resources of the pharmaceutical industry were directed towards the north star of profit rather than innovation. This could produce egregious examples of 'innovation' that helped shareholders far more than patients, such as in the glut of new cephalosporins, descendants of Brotzu's mould juice from Cagliari. Pharmaceutical firms could easily manufacture and then patent slightly different cephalosporins, though most offered no real medical improvements. They were what industry insiders call 'me-too' drugs.

From the 1980s onwards, antibiotics were no longer at the cutting edge of pharmaceutical development. It appeared that the low-hanging fruit had all been plucked in the golden age between 1940 and 1970. Soil-screening, once the powerhouse of antibiotic discovery, became a maddeningly repetitive exercise that turned up old antibiotics again and again. The rarity of antibiotics in soil matched their order of first discovery. Streptomycin, made by one in a hundred soil species, had been discovered in the 1940s by Albert Schatz working alone in a basement. In the 1950s it had taken an army of scientists at Lilly looking at soil from all over the world to discover vancomycin, made by 1 in 100,000. Lilly had continued screening soil where other companies stopped, which was why in the 1960s they discovered daptomycin, made by 1 in 10 million species. But daptomycin was only developed in the 1980s in response to growing vancomycin resistance. It was the last truly new antibiotic rather than a modification of an existing class, discovered in a bacterium isolated from the soil of Mount Ararat in Turkey, the site where Noah's Ark is said to have come to rest. Appropriately, daptomycin marked the end of the antibiotic flood.

The glacial pace of daptomycin's development after its discovery also showed the lengthy timescales that were now needed to bring a new antibiotic to market. Gone were the 1950s, when Pfizer had discovered Terramycin and got it approved in less than a year. The process of drug approval by the FDA had become more rigorous, and a new antibiotic had to pass through a daunting gauntlet of clinical trials: Phase 1 to show safety in humans, Phase 2 to show efficacy, and Phase 3 to prove that it was better than existing treatments. Negative results at any phase could return a product to the drawing board. Daptomycin was only approved as an antibiotic in 2003, sixteen years after its discovery. That was only slightly longer than usual: most antibiotics took between ten and fifteen years to become approved medicines.

At the start of the 1990s, the pharmaceutical industry attempted to tackle the slowdown in novelty by pinning their hopes on genetics. Now able to sequence genomes, companies identified all the proteins that were seen in bacteria but not mammals – deep genetic differences that trace back to our split over a billion years ago. Then, by testing for molecules that could bind to those proteins, they would develop entirely new broad-spectrum antibiotics. It was an ingenious approach – in theory.

Companies abandoned molecules made by microbes and instead screened their own libraries of chemicals, aided by advances in computing power. One firm's experience was indicative: from 1995, GSK spent seven years investigating 300 potential bacterial targets. Each 'screen' of a target involved running tests on over a quarter of a million possible compounds. GSK completed seventy screens, but only sixteen found a hit, and only five of those hits could be taken forward for more research. In the end, none of them became an antibiotic. By 2004, over thirty companies had published more than a hundred reports of screening programmes with similarly depressing results.

The dearth of new antibiotics meant that medicine was now entering a vulnerability window, a time when increases in levels and forms of resistance in bacteria would not be compensated for by new antibiotics. As one microbiologist reflecting on the history of beta-lactams put it, 'all the signs are that the fifty-year bonanza is well and truly over'. And the most extreme possibilities started to materialise, even more extreme than MRSA. In 2004, scientists in New York reported an alarming outbreak of multi-resistant bacteria in patients across seven different hospitals. These infections would normally have been treated with beta-lactam antibiotics, turning to different variants if one failed. But here, the bacteria were resistant to them all, including the carbapenems, the most recently introduced. These resistant bacteria became known as carbapenem-resistant *Enterobacterales* (CRE). Using beta-lactams in combination with an inhibitor, as in the case of Augmentin, wasn't an option for CRE either, because the bacteria carried an enzyme that could resist the inhibitor.

Like the exhausted defenders of a besieged city, the doctors had run out of ammunition. With few effective new antibiotics on the horizon for CRE infections, they turned to an old substance first discovered in 1949, an antibiotic called colistin. Colistin was so toxic and unpopular that it had been almost entirely abandoned for human use after its discovery, and consequently, CRE in the early 2000s had no resistance to it. But in an alarmingly short time, even colistin would be threatened by the emergence of resistance – not because of its use in hospitals, but because of its use on farms. The story revealed just how pervasively the modern world had come to rely on antibiotics.

Humans are omnivorous. Like our closest relatives the chimpanzees, we can survive without meat, but around 2.6 million years ago, our hominin ancestors began to increase the amount of meat they consumed. Evidence from archaeology suggests that consumption in prehistory was variable, probably driven by both cultural practices and local availability. But in modernity, everything changed. If global meat consumption over the span of the human species was plotted from 2.6 million years ago to the present, the most notable change would occur after 1950. During the second half of the twentieth century, the amount of meat eaten per person approximately doubled. The average global citizen now consumes over 40 kg a year, the average American three times that. This precipitous increase was brought about, in part, by antibiotics.

On Christmas Day 1948, a chance discovery revealed the role that antibiotics could play in creating a new and decadent world: a utopia where technology would allow the production of meat at a lower cost than ever before. Thomas Jukes was an American biochemist working at Lederle, the manufacturers of Aureomycin – one of the new tetracycline antibiotics. The massive increase in production of Aureomycin was also producing massive quantities of waste, and after brewing up gigantic vats of *Streptomyces aureofaciens* and extracting the antibiotic the company was left with the sloppy dregs to deal with. Currently these were being dumped into a nearby lake – but finding a way to use this apparently useless waste would be a pleasing efficiency. Jukes was experimenting with the dregs as a food supplement.

At the time, agriculture had become obsessed with optimisation. The growth of an animal was, in theory, a reproducible process. The farm would be treated like a factory: energy in, meat out. Rather than being kept outside, animals were increasingly housed under cover, reducing the amount of energy that was 'wasted' in movement. Each animal could then be fed *exactly* the same amount of food, laced with supplements that would enhance growth. There was great interest in a mysterious 'animal protein factor'. It turned out that animals grew faster when they were fed slightly carnivorous diets, such as fishmeal mixed into vegetable feed. In 1946, vitamin B12 had been identified as the molecule responsible. Humans rely on animal products for B12, which is why the vitamin is a common supplement taken by vegans and vegetarians. But although animal meat contains it, animals cannot make it. Every molecule of B12 ultimately originates from bacteria.

Merck, one of Lederle's rivals, had found that it could extract pure B12 from the bacterial waste left over from manufacturing streptomycin, which it could then sell to other companies. Jukes bought some of Merck's B12 and confirmed that giving it to chicks made them grow faster: they put on more weight in the same amount of time. But why buy from Merck when Lederle also had vats of antibiotic waste? He knew from testing that this waste also contained B12 – hardly surprising, given that Merck's and Lederle's antibiotics were produced by related bacteria. Jukes started a simple experiment, taking newly hatched chicks and separating them into different treatment groups. Some received Merck's pure B12 and some the unprocessed antibiotic waste in the form of a mash. A control group was fed a basic diet without supplements. Christmas Day 1948 marked the twenty-fifth day since the chicks had hatched, and the end of his experiment. If B12 was truly the secret ingredient, it would be reasonable to expect that the antibiotic waste would also lead to an increase in growth, but not as much as the more expensive pure B12.

The results showed quite the opposite: the chicks that had eaten the antibiotic waste had outgrown the competition. Those receiving the largest amount now weighed a third more than those that had received Merck's pure B12 – and an incredible 2.5 times more than the chicks that had received the basic diet. The antibiotic waste clearly had something extra that was responsible for even further growth. Jukes quickly established that it was the antibiotic itself. The extraction process was imperfect, and small amounts of Aureomycin were left in the waste. For some mysterious reason, Lederle's blockbuster antibiotic was also a turbocharger for chicken growth. At this point, Aureomycin was in such demand that there wasn't any pure antibiotic available for Jukes and his experiments. He resorted to visiting the company dump and scavenging for antibiotic dregs on discarded fermentation vessels.

Aureomycin was tested in other animals: it worked on turkey chicks as well as pigs, where it tripled their growth rates. That year, Lederle filed a patent for adding Aureomycin as a supplement to animal feed. The leftover liquid from antibiotic fermentation was pumped into railroad tank cars where it could be sent around the country. Word of this lucrative application got out. A story circulated about a pharmacist in the town where the processed pork meat SPAM was made who diverted a shipment of antibiotic residue, then sold it on the black market. He made so much money he retired to Florida.

It wasn't just Aureomycin that made animals grow faster. Other antibiotics had a similar effect, accelerating the weight gain of farmed animals like gasoline added to a fire. Scientists would go on to debate the precise mechanism: did antibiotics increase the amount of nutrition that went to the animal rather than its resident gut bacteria, reduce the thickness of the gut wall, inhibit low-level infections, reduce inflammation – or all of the above? But the immediate results were indisputable. In 1950, the *New York Times* told its readers that the discovery held 'enormous long-range significance for the survival of the human race in a world of dwindling resources and expanding populations'. In America in 1951, the FDA approved Aureomycin and five other antibiotics (including penicillin) as growth promoters for animals. The logic of antibiotic prophylaxis – giving antibiotics in the absence of illness – was applied to animals too. These combined effects meant that antibiotics became a 'universal lubricant' that was seen as helping agriculture run smoothly. The *Washington Post* reported that water laced with tetracyclines helped get chickens ready for slaughter in three-quarters of the usual time. An advertisement placed by Pfizer in *Time* told readers that 'platoons of little pigs were enjoying a peril-free infancy' – thanks to milk that had been spiked with Terramycin. Another from Cyanamid (Lederle's parent company) showed a chicken brandishing a bowl of antibiotics to ward off disease: 'Feed AUREOMYCIN Chlortetracycline to chickens and turkeys *continuously* at HIGH LEVELS . . . Give them *internal* sanitation'. The ad promised that, as a result, profits would be higher.

It seemed that there was no aspect of life, whether human or animal, that could not be enhanced by antibiotics. The pointless inefficiencies of the past that had held back economic growth could be circumvented using modern technology. Where the US led, the rest of the world wanted to follow: by 1955, antibiotic growth promoters had been licensed for use without prescription in West Germany, Britain, the Netherlands and France. From the other side of the Iron Curtain, the Soviet Union looked on with envy.

Nikita Khrushchev, born in 1894 to a poor peasant family in a small village in western Russia, had a life that mirrored the broader transition of his country from farming to heavy industry. He worked herding sheep and cows until the age of fifteen, before training as a metal fitter and working in mines and factories, rising through the ranks of the Communist Party to eventually become the Premier of the Soviet Union in 1957. 'I started working as soon as I could walk,' he once claimed. He didn't look like a typical politician: thick-set with broad shoulders, his youthful brawn only slightly softened into a generous paunch. Unlike many politicians, Khrushchev was under no illusions about rural life; he knew its deprivations and hardships first-hand, and he believed that technology had the power to improve them. He was unafraid of looking for new ideas beyond the Soviet Union, including to its main geopolitical rival. Previous Soviet visitors to America had informed Khrushchev of the wonders of its agriculture, assuring him that it was 'the most advanced in the world'.

In September 1959, when Khrushchev made a historic ten-day state visit to the US, the first ever such visit from a Soviet leader, he made sure that he took in its agricultural powerhouse: the American Midwest with its rolling cornfields. On the day he visited the state of Iowa to tour a local farm, Khrushchev was followed everywhere he went by a heaving scrum of 3,000 people. The corn farmer hosting him, Roswell Garst, threw silage at the massed journalists to encourage them to keep their distance. Khrushchev and Garst had met a few years earlier, when Garst had himself visited the Soviet Union. While there, Garst had told Khrushchev that corn was best used not as a cereal grain for human consumption but as animal feed: corn, he had explained to Khrushchev, was 'little sausages'. Khrushchev loved the farmer's metaphor and took to using it himself: 'corn is beefsteak, it is bacon, it is butter,' he said. 'That's why I am a booster for corn.' Khrushchev admired the American attitude to efficiency in agriculture. He saw no reason why capitalist methods could not be borrowed for socialist ends.

From his childhood, Khrushchev had always been fond of pigs. After visiting Garst's cornfields, he went to the Swine Research Center of Iowa State University. 'There are many cases', he told his hosts through a translator, 'when a swine is a more noble animal than even a human being.' (Some said he resembled one himself: a popular Soviet joke was a picture of Khrushchev in a pigpen with the caption 'Comrade Khrushchev, third from left'.) Khrushchev's hosts presented him with a model pig engraved with an awful pun: 'in the mutual interests of swi-ence'. Raising the pig aloft before the crowd, Khrushchev praised it as a 'fine American pig' – but he noted mischievously that it also had characteristics of a Soviet pig. Yet he knew all too well that in pork, America led the world – during WWII, the US had sent more pigs overseas than soldiers – aided by the quality of its breeding stock. The typical pig in the Soviet Union at the time was, in the words of one historian, 'petite, shaggy, and slow to reproduce'. Soviet farms were consolidated together to master economies of scale, attempts were made to breed a 'super-pig', but still the US remained the world's largest pork producer.

As Khrushchev walked past detailed displays illustrating the American scientific approach to rearing pigs, he expressed great interest. Their life cycle was divided into periods, each with its own calibrated rations. The final exhibit was a long line of all the components that were fed to an American pig. Twenty-four separate ingredients were laid out, from vitamin A to zinc – but the last, perhaps the most important, was antibiotics. An eleven-year-old local boy who met Khrushchev was unafraid of conveying American confidence in their superior methods. The boy told him that although the Russians might have recently landed the first probe on the moon just a few days earlier, America's strengths lay elsewhere: 'we can beat you in sausages'.

Intensifying pig agriculture was, as Khrushchev knew, a challenging problem. Pigs are foragers and roamers by nature. They enjoy company in small doses, tending to form groups made up of a few adult sows with their children, perhaps with a few single male pigs on the periphery. They are not herd animals like cattle, and when crammed together, their small lungs make them vulnerable to respiratory diseases. Previous efforts had been made to farm pigs more efficiently. In 1894, the year Khrushchev was born, a

Hungarian effort created a gigantic fattening plant on the outskirts of Budapest that housed over half a million of them. The ambitious plan failed when an epidemic broke out. Now, by protecting against disease, antibiotics loosened the main constraint that had thwarted massive pig farms. Mixing antibiotics into pig feed became standard.

The Soviet Union was already producing its own versions of American antibiotics – allegedly stolen through industrial espionage – and though they were often impure, they were used in all the same ways. New antibiotic plants were already being constructed to produce combined B12-antibiotic for livestock feed. But after Khrushchev's return, *Pravda* told its readers that production was now to be improved further by feeding animals 'bone and fish meal, fodder dregs, antibiotics, biostimulators, acidophilous preparations, and vitamins'. As the historian Claas Kirchhelle puts it, Cold War promises of development and prosperity created 'antibiotic proliferation'. Whether capitalist or communist, the world was locked in an antibiotic arms race: no single farmer, company or nation state wanted to be left behind.

With astonishing rapidity, antibiotics became a crucial cog in a vast new agricultural machine: factory farms. Pigs and other animals became trapped in a feedback loop. A desire for cheap meat justified using antibiotics and cramped conditions that were ideal for bacterial diseases to spread quickly – which in turn meant more and more reliance on antibiotics. In the name of efficiency, by the mid-1960s, American swine breeders were experimenting with packing sows into metal crates. Hemmed in by bars, a sow was unable to turn around – the only actions possible were eating the food in front when it appeared and defecating behind, the manure sluiced out into a lagoon of slurry. Confining pigs reduced the energy that they would otherwise waste by moving around. The traditional world of a pig – fields, sky and sun – was no more. The modern farm shrank their existence in every sense. Animals lived faster and died younger without ever knowing the outside world, slaughtered on the altar of economic growth.

Despite these developments, throughout the twentieth century pigs remained challenging animals to farm for meat. They took longer to mature than poultry, and unlike cattle they didn't produce useful milk while they were growing. Combined with short pregnancies and large litters, the pork industry was especially prone to cycles of boom and bust driven by fluctuating prices. Antibiotics helped ease these difficulties and avoid the sudden loss of whole sheds of animals due to disease. Across the world, there was local variation in which antibiotics were used, and how, but many of them were also being used in human medicine. Some, however, were almost exclusively agricultural. One such group was the polymyxins, which had first been discovered in the late 1940s.

Much like the tetracyclines, the polymyxins had been discovered multiple times: 1947 had seen three publications in quick succession from two American laboratories and one in England, followed by an independent discovery in Japan in 1950. The Japanese discovery was made at a laboratory that had been set up within a company that manufactured soap. Appropriately enough, the antibiotic they discovered – colistin – worked like a detergent against Gram-negative bacteria. Bacterial cell membranes are made of lipids: fatty molecules that repel water. Gram-negative bacteria have an outer membrane containing a dense layer of lipids, staunchly protecting its interior from foreign molecules. But a molecule of colistin weakens the lipid membrane, like detergent breaking apart a clump of dirt. The membrane becomes patchy, increasing the cell's permeability to water; quickly, water flows into the cell and it bursts. Unlike the tetracyclines, which were given to both humans and animals, colistin was rarely used in hospitals because it was unacceptably toxic. Lipids are a component of the bacterial membrane, but also an important part of human cells. When colistin circulated in the bloodstream and passed through the kidneys it was filtered out of the blood. That meant it silted up to high concentrations, to the point where it acted like a detergent on human cells too: a molecule meant to destroy bacteria could also destroy the kidneys. This toxic side-effect happened in around 40 per cent of patients treated with colistin.

After its discovery, colistin was used a little to treat penicillin-resistant infections of Gram-negative bacteria. But soon, much better antibiotics were developed for the same purpose; it became unnecessary to risk colistin's toxic side-effects. As time passed, colistin became an embarrassing legacy from a less sophisticated age, like a boorish elderly relative at a family party. But like many other antibiotics, low levels of colistin in animal feed worked as a growth promoter. Which antibiotics got used where for growth promotion was often arbitrary and parochial; in the US, colistin was almost never used in agriculture. But elsewhere, it was.

One of those places was China. From at least the 1980s, farmers in East Asia were giving colistin to piglets. For other antibiotics that were important in human medicine, agricultural use might be dangerous: there was a risk of breeding resistance that could spill back into the human population. By contrast, using a retired antibiotic like colistin seemed harmless – the likelihood it would ever be needed again seemed extraordinarily low. It was simply an old drug gathering dust. But in the mid-2000s, as bacteria like CRE made their unwelcome appearance in hospitals around the world and new antibiotics were in short supply, doctors found themselves reaching far back in the medicine cabinet, far back enough that, reluctantly, their fingers brushed up against colistin. There was nothing else left.

Doctors turned back to colistin as a last resort. It was so toxic that it would have been unlikely to pass the safety studies required of a new antibiotic, but because it had been approved before such studies were needed, it enjoyed 'grandfather' rights. For some doctors, it was only ethically acceptable to give colistin to a patient when the alternative was administering last rites. One patient described the experience as 'like my organs were disintegrating. Like I was dying from the inside out.'

Varying patterns of antibiotic use fanned the flames of antibiotic resistance higher in some places than others. In Greece, where beta-lactam use was extremely high, rates of deadly CRE infections in the mid-2000s rose far more quickly than in many other countries. Doctors were forced to turn to colistin, with devastating effect: soon they had bred infections that could resist it. The rise of colistin resistance in Greece was dramatic: from nothing in 2007 to 8 per cent of CRE infections in 2008, it reached 24 per cent in 2009 and kept on rising, climbing to 40 per cent in 2016. Fortunately, colistin resistance remained largely concentrated only in the hospitals where colistin was being used. Indeed, one of the few advantages of colistin for doctors was that colistin resistance had never been observed on a plasmid, so its spread was limited. It was tempting to believe that because the drug targeted something so fundamental – the outer membrane of the cell – the only way to acquire resistance was to alter multiple basic genes in the chromosome by mutation. Unlike a plasmid carrying a resistance gene, evolutionary theory suggested these damaging mutations would disappear again once colistin wasn't used. This wishful thinking didn't consider that the increase in use of colistin to treat CRE infections was tiny compared to its main use worldwide: as a growth promoter in agriculture.

In the European Union alone, the average consumption of colistin and other polymyxins was over 300 times higher in animals compared to human medicine. Concerned about the risk of antibiotic resistance in animals spilling over into humans, the EU had enacted a complete ban of antibiotics as growth promoters in 2005. But their use was still permitted for other purposes – and in the view of some critics, that loophole allowed growth promotion to continue under the guise of antibiotic prophylaxis. Colistin was an important antibiotic for farmers: a survey found that the polymyxins were the fifth-most common antibiotics sold for animal use. By 2015, an estimated 12,000 tonnes of colistin were being used worldwide every year. The majority, 8,000 tonnes, was being used in China – entirely in animals.

In the mid-1970s, the average Chinese citizen ate just 8 kg of pork a year, almost all of it produced from small backyard farms. The industrialisation that Khrushchev had been so eager to effect in the USSR twenty years before as yet had no parallel in China. It was only in the 1980s, inspired by the examples of the US and the Soviet Union, that the Chinese government began in earnest to standardise pig breeds, regularise their food and harmonise the medicines they were given. China's economic growth over subsequent decades created an expanding middle class eager to eat pork. By the turn of the twenty-first century, the Chinese government was promoting large factory farms to accelerate the agricultural transition. The state had even created its own secret pork reserves to weather fluctuations in the notoriously volatile global pork market, maintaining warehouses filled with unknown amounts of frozen pork and live pigs. By the 2010s, Chinese citizens ate more pork per capita than Americans. The dramatic expansion of Chinese pig farming was so spectacular it was dubbed the 'pork miracle': by 2015, half of all the world's pigs lived in China. But calling it a miracle was to understate the deliberateness of the process. One side-effect of the drive for efficiency was Chinese agriculture's enormous consumption of colistin. And by the end of the same year, China's colistin use made headlines.

In November 2015, Chinese researchers published a shocking discovery. The tonnes of colistin that had been poured into the nation's agricultural machine had summoned a new resistance gene that was now spreading into bacteria that caused human infections. Colistin, an antibiotic of last resort, was under threat.

Earlier in 2015, the British scientist Tim Walsh had been visiting Beijing for an ongoing collaboration with scientists at China Agricultural University. Walsh had made a name for himself as a chronicler of the spread of new resistance genes around the world. He was fascinated and horrified by the seemingly limitless way that they could emerge in hospitals and then spread globally. Working on new and alarming resistance genes in CRE, he had wondered in 2005 whether these early observations were 'the quiet before the storm'. The visit to China had been full of scientific discussion, and when Walsh left for the airport, his colleague Yang Wang came with him in the taxi to continue the conversation. In the taxi, Wang told Walsh about a new finding, one so secretive that he seemed to have waited until this moment to let it slip: while studying bacteria from Chinese pigs, he and his colleagues had discovered a new form of resistance to colistin that was carried on a plasmid. Walsh was so shocked he cancelled his flight and turned the taxi around.

Colistin resistance on a plasmid was supposed to be impossible. A scientific journal had rejected the first draft of the paper, finding the results implausible and unsupported. Walsh pressed Wang for more details. Walsh knew the scientific story was massive because of the unprecedented nature of the resistance; he desperately wanted to be involved in the paper, even at this late stage. In his career, Walsh had shown himself to be an assertive, even abrasive operator who often refused to take no for an answer. A few years earlier he had become embroiled in a diplomatic row with the Indian government about antibiotic resistance so extreme that he was warned not to travel to the country. He insisted he could help the Chinese scientists get the paper published, and with Wang vouching for him he was allowed into their project. They did some further experiments, rewrote the paper and had it published by November.

The reaction was seismic. Newspaper headlines proclaimed that the world was on the cusp of a 'post-antibiotic era'. The new resistance gene encoded an enzyme which the scientists called mobile colistin resistance (MCR-1), a grim confirmation that bacterial adaptability could transcend what scientists thought possible. As a resistance enzyme, MCR-1 was fascinating. It directly modified lipids in the cell's outer membrane, making colistin's binding less effective and allowing the bacteria to tolerate higher levels of the antibiotic. The gene appeared to have originated in bacteria called *Moraxella* that could be found in the respiratory tract of pigs. It seemed a random shuffling of the gene out of *Moraxella* onto a plasmid, combined with the presence of colistin, had allowed the plasmid to spread MCR-1 between other bacterial species present in pigs – and then into people.

The first observation of MCR-1 reported in the paper had come from a bacterium isolated from a Chinese pig in 2013, but that chance observation was the tip of the iceberg. How long had it been spreading? Using old samples of bacteria stored in laboratory freezers allowed scientists to perform historical reconstructions, like lexicographers looking through old newspapers for the first occurrences of a word. The earliest observations of MCR-1 were from chickens in the 1980s, which matched when colistin had first started to be used in food-producing animals in China. After that, MCR-1 was undetectable until the mid-2000s, at which point it reappeared and began to rise, matching the heavy increase in colistin use during the Chinese pork miracle. By 2016, the gene was present in an incredible 45 per cent of *E. coli* on pig farms. Concurrently, the prevalence of MCR-1 in patients in Chinese hospitals had started climbing. At less than 1 per cent in 2011, it grew to an alarming monthly peak of nearly 50 per cent in November 2016. Colistin wasn't used in the Chinese healthcare system: this was *all* due to spillover from agriculture. There were any number of possible routes for the gene into humans, and it was difficult to disentangle them. Bacteria could be smuggled inside the pork and unknowingly consumed; contaminated manure could flow into water systems; insects or birds could pick up bacteria from farms and fly them into homes and hospitals. The boundaries were porous.

The spread of MCR-1 was not simply a Chinese problem. The logbooks of freezers showed its spread around the world: it had been present in Japanese pigs in 2007, Spanish pigs in 2009, German pigs in 2010, then French and Belgian pigs in 2011. In 2012, it showed up in pigs in Vietnam and Laos, and in American pigs in 2016. The global trade in live pigs could easily move bacteria around the world. But MCR-1 wasn't only confined to pigs. It had been detected in bacteria from guinea fowl pie in France, a boiled potato in Bolivia and vegetables in Switzerland. In Germany, researchers found it in bacteria from grass lizards that had been imported from Vietnam. MCR-1 was even seen in a bacterium from the foot of a penguin that had washed up malnourished on a Brazilian beach, having migrated north from the shores of Patagonia in search of fish shoals. MCR-1, once an unassuming enzyme in an environmental bacterium, appeared to have spread all over the world in just a few years.

Although nobody could be definitive about where MCR-1 had first emerged, evidence suggested that China was the most likely place. The spread of pathogens between humans and animals had sparked a global fire; the similarities with viral pandemics like SARS (and, later, Covid-19) did not go unnoticed. Diplomatically, the origin of MCR-1 created an embarrassing situation for the Chinese government, which wanted to be seen to act quickly. In July 2016, the Ministry of Agriculture announced a ban on colistin's use as a growth promoter. The results of the ban were dramatic: production of the premixed form of colistin used in feed in China decreased from over 27,000 tonnes in 2015 to less than 2,500 tonnes three years later. That fall in colistin usage almost immediately reduced the prevalence of MCR-1, with the proportion of resistant bacteria detected falling in both agriculture and humans.

The Chinese ban was followed by other countries. Brazil, a huge poultry producer, banned colistin as a feed additive in November 2016. Other countries were hot on its heels: Thailand, Japan, then Malaysia, Argentina and, by July 2019, India. However, colistin didn't go away. Tim Walsh worked with collaborators in Pakistan, Bangladesh and Nigeria to map out the international trade in colistin. Their study was a cross between a piece of investigative journalism and a scientific paper, and it showed that countries where colistin had been banned were continuing to export it to other countries where it was still being used. Colistin was made in China, sent to intermediate companies in Europe, then shipped onwards to countries like Pakistan, Nigeria, Kenya and Bolivia. Even though its application in agriculture had reduced, poultry and pigs still accounted for 96 per cent of colistin's global use. And though the bans did reduce levels of MCR-1, they did not eliminate it. Like other resistance genes, MCR-1 was now out of the bottle.

Colistin resistance wasn't the end of the world, but it was an ominous glimpse of a dangerous future. The world that antibiotics had made was unsustainable. In the second half of the twentieth century, antibiotics had transformed society far beyond medicine through their role as growth promoters for animals. Step by step, the whole world had turned to them to increase agricultural efficiency.

Their deployment wasn't a simple linear story: superpowers like the US, the Soviet Union and China had all, at different stages and in their own ways, tried to intensify agriculture, relying on huge supplies of antibiotics. Some countries, like those in the EU, had belatedly taken steps to decrease antibiotic consumption in animals, but only after benefiting from its efficiencies. This made it hard for them to make a convincing moral argument against their use elsewhere, much like post-industrial states urging developing nations to curb their fossil fuel consumption. Though colistin might have been banned for growth promotion in China, the intensification of farming did not end. After the ban, gargantuan new pig farms were still being constructed: high-rise blocks up to twenty-six stories high, where millions of pigs were separated into floors based on their life stage as if J. G. Ballard had designed a porcine dystopia. In a competitive world, the economic imperative for efficiency and the fear of falling behind were more powerful drivers than the hypothetical and marginal risks of breeding resistant bacteria that might eventually affect human health.

From one perspective, the MCR-1 tale was an optimistic one where governments had taken quick action to deal with the problem. Yet an inconsistent patchwork of regulations meant that the agricultural use of colistin continued. And not just colistin: many other antibiotics became seen as essential for modern farming, with global forecasts projecting a continued increase in their use.

The flow of bacteria from animals into humans is not necessarily a key driver of resistant infections; humans tend to fall sick with bacteria that come from other humans, unless they are in regular contact with animals. But colistin resistance confirmed that the boundaries between people and agriculture were porous. Like the earliest observations of MRSA in the 1960s, the detection of colistin resistance rang alarm bells across the world. Wherever new forms of antibiotic resistance emerged, they could spread, and what was seen first in agriculture could easily become endemic in humans. Transmission in the other direction had also been shown to be possible: in the mid-2000s, MRSA from humans passed into pig populations. Bacteria knew no borders.

No viable path

Plazomicin

Nature herself is a pretty good terrorist.

Hank Heine

San Francisco is a city built on microbes. The reddish-brown rocks of the headlands of the Bay Area were formed from the compressed skeletons of single-celled microbes, radiolaria, which lived, died and rained down silently on the ocean floor millions of years ago. As the Pacific tectonic plate collided with the North American, the seabed rose up and formed the precipitously steep hills around the Bay Area, meaning that, in time, the inhabitants of the wealthiest city in America would need to park their cars with a characteristic twist of the wheels into the curb. As the city boomed from the nineteenth century to the twenty-first, it was capital that accumulated in San Francisco.

Geography shaped where that capital settled. In the 1890s, a wealthy industrialist had chosen an area of wasteland south of San Francisco as the site for a new meat-packing factory, where the stench wouldn't bother the increasingly genteel city-dwellers. As the twentieth century marched on, the new industrial area that sprung up around these factories, South San Francisco, became a nexus for the steel industry and shipbuilding. By the mid-1960s, beatniks and dreamers were gathering in San Francisco's cafes and bookshops, a world away from the suburbs of South San Francisco just a few miles away, now filled with the decaying remnants of industries that had moved on: derelict warehouses, rusting steel mills and empty shipyards. But in 1976, when a new company called Genentech was looking for a cheap place to rent, it found a perfect semi-vacant warehouse just south of the old shipyards. The ebb and flow of capital had made and unmade South San Francisco; now, its undesirability meant that once again it would be reborn, this time as the birthplace of biotechnology.

Genentech had been founded by a biochemist at UC San Francisco together with an enthusiastic venture capitalist. Its hope was to use recombinant technology, learned from bacterial plasmids, to manufacture human insulin. Insulin is a natural hormone made in the pancreas which regulates blood sugar, but people with type 1 diabetes do not produce it and will die without regular injections. In 1976, the only way to obtain insulin was as a byproduct of the meat-packing industry that had first made South San Francisco's fortunes. Harvesting one pound of insulin required the pancreases from over 20,000 animal carcasses.

Genentech hoped to use bacteria to make insulin, just as microbes had been increasingly manipulated into manufacturing antibiotics. But first, bacteria needed to be given the gene. The scientists inserted genes they had synthesised onto a plasmid and then placed it inside *E. coli*, whereupon the bacteria dutifully started producing insulin, as pure as if it had been made in a real human pancreas. These genetically modified *E. coli* made so much insulin that through a microscope they looked like stuffed olives. The company became similarly stuffed with capital after it licensed its technology to pharmaceutical giant Eli Lilly, swelling from a few people in an old warehouse into a company employing thousands. By the end of the twentieth century, the Genentech headquarters on Oyster Point resembled the campus of a private university, with over forty buildings arranged around a curving central road: DNA Way.

At the beginning of the twenty-first century, humanity was able to manipulate microbes in increasingly sophisticated ways. Yet companies had become less and less interested in antibiotics, attracted elsewhere by more lucrative opportunities. What would happen when the world needed a new wonder drug to beat the next bacterial threat?

In 2004, a small group of scientists and investors in San Francisco had noticed this growing gap in the market, a gap through which they hoped to slip. If they could anticipate the future of bacterial evolution and develop antibiotics that worked against the deadliest drug-resistant strains, they would have a product ready to go when the crisis arrived. By investing capital in the future of antibiotics at a time when others were putting their money elsewhere, this small group was like the nineteenth-century meat-packing industrialists, hoping to transform what others saw as a wasteland into great riches.

The company was a startup called Achaogen. It had been co-founded out of the Scripps Research Institute in La Jolla, California, by a group including PhD student Ryan Cirz. Its CEO was Kevin Judice, a chemist who like many in South San Francisco had witnessed the decline in antibiotic research first-hand. When Judice had started working on antibiotics to combat MRSA at another company in 1997, the first couple of international meetings he'd gone to were well attended and vibrant. But after that, each year the talks were less upbeat and fewer people were listening. The conference rooms were emptying as the great flow of capital needed to power new antibiotics dried up.

Achaogen was entering what looked like a dying field. Despite the widely publicised alarm over the rise in resistant bacteria like MRSA and CRE, large pharmaceutical companies at the time were less interested in pursuing antibiotics. Although CRE were far from a public health crisis in the US, having been seen only in a handful of unfortunate patients, Judice thought the example of MRSA could be instructive. If CRE continued to rise in prevalence, then in time there could be a huge market for new antibiotics to treat them. The Gram-negative CRE presented even more of a challenge than the Gram-positive MRSA. When Hans Gram had found that dye molecules did not soak into such bacteria in the 1880s, he had not known the reason. In the 1960s, researchers had discovered that Gram-negative bacteria had an additional outer membrane around their cells that prevented the dye molecule getting through. (It was this outer membrane that colistin attacked.) Antibiotic development against Gram-negatives was harder for this reason: it was a struggle to find molecules that could get inside. When Judice looked at the absence of other companies investing in new Gram-negative antibiotics, he felt that the terrifying scenario was also a potentially profitable opportunity.

Judice was fond of quoting a saying attributed to James Black, a British chemist: 'If you want to make a new drug, start with an old drug.' His plan followed the blueprint of Sheehan's vision for the penicillins: Achaogen would choose an existing antibiotic, known to be effective and safe in humans, then deliberately modify it to make it invulnerable to current forms of resistance. Judice recruited a Swiss chemist called Heinz Moser who had worked at the pharmaceutical giant Novartis. It was decided that Achaogen would try to make a new version not of a penicillin, but another of the oldest antibiotic classes: the aminoglycosides. This class had a strong pedigree – streptomycin was the first aminoglycoside – but had fallen out of favour since the 1970s. Achaogen planned to bring them back with a vengeance.

This was not a trivial undertaking. Aminoglycoside chemistry was hard. On paper, the conventional diagram of an aminoglycoside molecule looked floppy and flexible, but in reality, its atoms formed lots of weak bonds, giving it more structure than expected. Modifying any part of the molecule could disrupt the rest of the structure and destroy its antibiotic activity. After deciding to focus on the aminoglycosides, Achaogen recruited its secret weapon. George Miller, in his late sixties with bushy white eyebrows, was the oldest employee at the new startup by far: officially his title was Research Fellow. His colleagues felt he knew more than anyone else alive about aminoglycoside resistance. Achaogen also recruited a younger chemist called Jim Aggen to work with Miller and perform experiments.

Their plan started from a natural aminoglycoside. Two hundred miles inland from Achaogen's headquarters, the other side of the jagged Sierra Nevada, lay Inyo National Forest. In 1970, researchers had isolated a new species of *Micromonospora* from Inyo soil and found that it made a new antibiotic called sisomicin. Bacteria could become resistant to sisomicin and other aminoglycosides by acquiring an enzyme that modified the molecule at particular points, slotting in like a key into a lock, turning the antibiotic into a dud. If the sisomicin molecule could be chemically altered at these points to start with, the key would no longer fit. The difficulty was to stop the key fitting while making sure that the modified molecule still worked as an antibiotic.

It was a delicate balancing act: changing a molecule while keeping it the same. When Aggen needed to let off some steam, he would go on a walk from Achaogen's headquarters in South San Francisco. Achaogen's building was just off the main highway, Route 101, right opposite San Bruno Mountain. A gigantic sign in white concrete letters on the hillside welcomed visitors to

SOUTH
SAN FRANCISCO
THE INDUSTRIAL CITY

San Bruno was not an attractive place to go hiking. The city dump was over the other side of the hill. Often, the incoming wind off the Pacific would pick up trash and blow it over the top in a steady rain of detritus. Also, about once a year the whole hillside would catch fire and burn down to the freeway. Looking out from his office, Judice would sometimes see Aggen trudging up the mountain by himself, thinking about aminoglycosides.

Achaogen was a small company. As CEO, Judice needed to ensure that investment kept flowing in to fund its research and development. It was difficult to persuade investors to put their money into a vague future crisis that might not come to pass – particularly in the fearful and neurotic world of post-9/11 America, where antibiotic resistance was far down the list of existential threats. But here Judice saw another potential opening.

In the weeks following 11 September 2001, the US had been gripped with fear as letters filled with white powder began to arrive at news organisations and the offices of politicians. The powder was spores of *Bacillus anthracis*, commonly known as anthrax. These tiny spores, the analogue of those formed by *Penicillium* that had landed in Fleming's Petri dish, were deadly. Anthrax causes several thousand infections a year globally, mostly skin infections in people who work with cattle. But if its spores are inhaled, they can be a deadly weapon – a tiny puff of them would be enough to kill. In the late 1970s, an accident at a Soviet bioweapons facility led to a cloud of anthrax spores rising up into the air, where they were carried gently on a northerly wind. At least 105 people died, the activated spores moving from their lungs into their lymph nodes, then gradually germinating and blooming throughout the body, releasing toxins that damaged tissues and caused respiratory arrest. Autopsies of anthrax victims showed the lining of the brain filled with dark red blood, like a cardinal's cap.

Infections of inhalational anthrax were vanishingly rare. Only eighteen cases had been recorded in the US in the twentieth century. In 2001, nobody had died of the disease in over twenty-five years. But less than a month after 9/11, health officials in Florida announced to a nervous nation that Bob Stevens, a photo editor for a tabloid newspaper, had died of the disease.

Chaos ensued. In total, the anthrax letters killed five people, two of them postal workers, and made at least seventeen people sick. People became terrified of opening their mail, leaving their bills and mortgages unpaid. There was a run on antibiotics and gasmasks; Vice President Dick Cheney travelled everywhere with a hazmat suit. The cost of decontaminating buildings alone ran to $320 million, with the full economic impact of the letters estimated at several billion dollars. The culprit or culprits remained unknown, but they had inflicted one of the most efficient terrorist attacks in US history: by posting a few letters, they had paralysed an entire nation.

There was huge pressure on the government to do something. With no clear culprit, attention turned away from domestic terrorists and towards foreign states. When President George W. Bush gave the State of the Union Address in January 2003, he claimed that American citizens faced a very real threat from 'outlaw regimes' that possessed anthrax, singling out Iraq and President Saddam Hussein. When Colin Powell spoke to the United Nations Security Council the following month, he reminded them that 'less than a teaspoonful of dry anthrax in an envelope shut down the United States Senate' – adding darkly that Hussein could have enough anthrax for 'tens of thousands of teaspoons'.

After the US invasion in March 2003, it would become clear that Iraq had no anthrax. Yet fear proved a powerful motivator. Bush had promised that the US would launch a major research and production effort against bioterrorism. The plan was called Project Bioshield, and it mandated the government to maintain gigantic stockpiles of treatments to be used in the event of bioterrorism. When the Bioshield Act was passed into law in July 2004, the year Achaogen was founded, the US government committed to spending $5.6 billion over the next ten years, much of it funnelled through the US military.

The money was there and it needed to be spent. But on what? At a time when cash for antibiotic research was scarce and investors were wary, Judice wondered if the US government would want to fund Achaogen. After all, what if the next bioterrorist used antibiotic-resistant anthrax? There was little difference between naturally resistant bacteria and bioterrorism. His pitch was that funding new antibiotic development was a matter of national security. However, army generals were not accustomed to playing drug developers, and Judice didn't know how to persuade them. Interacting with the American military-industrial complex in all its might was something that, as a self-confessed leftwinger from the West Coast startup world, he found challenging. Fortunately, he had recruited a business development officer, John Hollway, whose father had served in the military, and who could act as a translator. In one meeting, Judice accidentally forgot to call someone 'General'. Hollway explained to him afterwards what a faux pas he'd made. 'I know what you've got is hot shit,' he told Judice. 'But they're not going to care if you don't act like you understand their system.'

Achaogen started to understand the system. In 2006, the Department of Defense awarded the company $24.6 million. That year, the Achaogen chemists had also made something of a breakthrough with their attempts to modify sisomicin. In total they had constructed over 400 variants of the molecule which they screened against resistant bacteria. The best molecule was one referred to at first only as ACHN-490. When they tested it against a range of multi-drug-resistant bacteria from hospitals, including the deadly CRE, its activity looked promising. It squared the circle: it retained the activity of aminoglycosides without being vulnerable to the same resistance mechanisms. The chemists dubbed their new molecule plazomicin.

Achaogen only got better at winning government funding. In June 2007, the Department of Defense awarded it a further $18.8 million. The company started a Phase 1 safety trial of plazomicin in February 2009. Then, in March, the National Institutes of Health gave it a $26.6 million contract. Project Bioshield was still running in 2010, and had set up a new funding agency called the Biomedical Advanced Research and Development Agency (BARDA). In August 2010, BARDA announced its first awardee of an initiative to help develop new broad-spectrum antibiotics: it was giving Achaogen a $64 million contract to fund plazomicin research. By July 2010 plazomicin was in Phase 2 trials, and Achaogen was 'flush with cash', in the words of one profile. Framing its research as a matter of national security had helped Achaogen raise nearly $150 million in research funding. Ironically, it began to have too much money. Once, it even tried to return some to the US government unspent; it was told that this wasn't appropriate.

By 2011, plazomicin had successfully passed Phase 2 trials. The next and final phase was perhaps the most daunting: proving that plazomicin worked better than existing treatments. Judice was feeling burnt out, and decided to hand over as CEO to a Scottish executive called Kenneth Hillan – another Bay Area scientist who had worked at Genentech.

Hillan faced a formidable task. Achaogen was about to try and convert itself from a research startup, in need of continuous injections of capital, to a pharmaceutical company that returned a profit.

Pharmaceutical companies are, like most of us, desperate for approval. For a new medicine to be approved by the FDA, its developer must specify its intended use so its efficacy can be assessed. The technical term for the drug's intended use is its 'indication'. However, once a drug has approval, its use isn't restricted – a doctor can choose, if they wish, to prescribe it for something it wasn't approved for in an 'off-label' prescription.

Companies often exploited this legal situation. One example was the enormously successful antibiotic ciprofloxacin. Patented by Bayer in 1980, it had received its first approval by the FDA in 1987 and was then heavily advertised to doctors as a broad-spectrum oral antibiotic. During the first six months of 1988, ciprofloxacin was the second-most advertised product in pharmaceutical journals. Unsurprisingly, it became one of the most widely prescribed oral antibiotics. Bayer heavily promoted the drug to doctors, but it had only marginal activity against some common bacteria. In the US, ciprofloxacin was bringing in huge profits for uses outside the indications that the FDA had approved it for. All of this was perfectly legal. In 1990, two doctors noted that Americans were spending $700,000 on ciprofloxacin – every day. 'Our experience', they wrote, 'suggests that much of this expenditure is inappropriate.'

The ease of winning approval for plazomicin would hinge on Achaogen's choice of indication. Ideally, they would choose a condition where existing treatments were bad to non-existent, so that plazomicin's superiority would be clear. The drug was intended to protect against the growing number of multi-drug-resistant bacteria like CRE, and so Achaogen decided to focus on CRE bloodstream infections – a cause of sepsis – which had a fatality rate of one in two. Achaogen was confident that plazomicin would outperform colistin, the notoriously poor antibiotic that until now had been the best available option.

The principle of the planned trial was simple: Achaogen would recruit patients with serious CRE infections – both bloodstream infections and pneumonia – and randomly treat them with either colistin or plazomicin. In practice, since patients were seriously ill in intensive care units and many were expected to die within days, the trial was a high-stakes enterprise. There was simply less time for all the usual paperwork that a modern trial requires – getting consent, screening for inclusion, assigning the patient to receive one drug or the other, monitoring them carefully – all while they were potentially dying of infection.

In recognition of plazomicin's potential importance, the FDA had awarded Achaogen a coveted Fast Track Designation, a bureaucratic stamp of approval that would speed up the trial's planning process. But designing the trial was complicated, because carbapenem-resistant infections like CRE were rare. In 2013 in the US, there were fewer than 3 such infections per 100,000 people. Even fewer of those represented bloodstream infections or pneumonia in hospitals.

The trial didn't begin recruiting patients until 2014. Hillan, Achaogen's CEO, told a Senate committee that the FDA had been 'extremely collaborative'. He said he hoped that Achaogen's experience could be a model for future antibiotic development. Within months, the FDA would confer yet another mark of distinction on plazomicin, guaranteeing Achaogen priority review of the results as well as an extra five years of exclusivity on plazomicin beyond the usual twenty-year patent. According to *Business Monitor*, Achaogen was now an attractive acquisition target for a big pharma company.

At the start of 2014, Achaogen went public. Up until that point, the startup had been funded by private investors and grants, but now its shares were trading on the stock exchange. The reason was simple: a need for capital. Achaogen's leadership hoped to expand the company with the extra investment, giving it the funds to run the trial and start manufacturing and selling plazomicin. Analysts recommended investing, on the basis that if plazomicin was successful there would be 'huge profits'.

But going public had risks as well as benefits. Being floated on the stock market meant that for the first time in Achaogen's history, people could bet against them.

When recruitment for the trial began, things started to go wrong. Achaogen had agreed with the FDA to recruit 360 patients, which it had been estimated would take three years. They were recruiting not only in the US, where CRE rates were low, but also in countries with higher infection rates – across Southern Europe into the Middle East, adding more sites in Latin America over time. But enrolment was slow. In consultation with the FDA, Achaogen amended the trial design to try to speed it up. Investigators at hospitals worldwide screened over 2,100 patients. By August 2016, two years after the trial started, they had recruited just sixty-nine patients.

Achaogen's endeavour had become an expensive disaster. The average patient spent just twelve days in the trial, but cost about $1 million to enrol. At the current rate, reaching the originally intended number of patients would take a decade and cost over $300 million; Achaogen had neither the time nor the money to see the original plan through. The company decided to cut its losses and prematurely end the trial.

Fortunately, as it had become clear how difficult the trial would be, Achaogen had hedged its bets and concurrently started another Phase 3 trial for a different indication. The company would test plazomicin against so-called 'complicated' urinary tract infections: any infection that didn't respond to initial antibiotic treatment or failed to resolve quickly. Though CRE could also cause these infections, so could many other bacteria. Urinary tract infections were much more common than those in the bloodstream and involved patients who weren't at death's door, making this new trial a far easier prospect. By December 2016, both Phase 3 trials had concluded. For the time being, their results were a closely guarded secret.

On Monday 12 December 2016, John Hollway – the business development officer who had helped Achaogen obtain military funding earmarked for bioterrorism – no longer worked for the company. But he still held shares. He was in Philadelphia, about to board a plane to Las Vegas for a work trip, when his brother-in-law called to tell him that Achaogen had published the Phase 3 results for both plazomicin trials.

The results from both trials were impressive. Elated, Hollway got on the plane. As he soared over America, Achaogen's shares were soaring on the stock market: by the time he'd landed, they had more than doubled in value. That evening in Vegas, he took himself out for a fancy meal on his own to celebrate. Afterwards, he was walking through the gold and marble forum of Caesars Palace when he realised he could now afford to splash out on a Christmas present for his wife. Impulsively, he walked into the Prada store and bought her a handbag. This was so out of Hollway's usual pattern of spending that his credit card company flagged it as a fraud attempt and contacted his wife. She phoned him up immediately. 'Either you're having an affair,' she said, 'or I just got something *really* nice for Christmas.'

Hollway wasn't alone in his optimism. Achaogen shares surged and analysts raised their estimates of how high the stock could go. It looked like a vindication of research into new antibiotics. It was a good time for antibiotics: the following day, President Obama signed the 21st Century Cures Act into law, which aimed to allow certain drugs to get approved more easily. The Act would either help life-saving treatments reach patients faster with less red tape (so said the pharmaceutical industry and their lobbyists) or allow substandard medicines to bypass the tried and tested requirements of clinical trials (so said the FDA). Among its many provisions, the Act created a special pathway to FDA approval for antibiotics. For cases where an antibiotic was meant to treat a 'serious infection in a limited population of patients with unmet clinical needs', the FDA could approve it for use in that limited population only. This route was meant to provide a way for antibiotics to get approved for infections where running clinical trials was exceptionally difficult. Cases, in other words, exactly like using plazomicin to treat life-threatening CRE bloodstream infections.

By March 2017, Achaogen's market value was over $1 billion. That year, the WHO published a list of priority bacteria to target for new antibiotics: CRE were given the highest possible status. Achaogen spent 2017 preparing its FDA application for plazomicin approval. In January 2018, the FDA confirmed that plazomicin for CRE bloodstream infections would be the first application reviewed under the new regulatory pathway. Everything looked promising. Achaogen had announced its plans to seek two indications at the same time: plazomicin for both bloodstream and urinary tract infections. It was a mark of its confidence in its new drug.

On the morning of 2 May, Achaogen representatives and FDA officials gathered in the largest room in the Doubletree Hotel in Bethesda, Maryland. The room could seat over 400, but the key people in it that morning were the fifteen independent academic and medical experts making up the FDA's Antimicrobial Drugs Advisory Committee. For the rest of the day, they would listen to arguments for and against the approval of plazomicin, and then they would vote, knowing that the FDA nearly always followed their recommendation. The outcome of the meeting would shape the flow of millions if not billions of dollars.

Outside, it was a warm summer's day, but the atmosphere in the hotel was chilly as Achaogen's Chief Medical Officer presented the results for CRE bloodstream infections. Comparing the two groups of patients, the death rate of those who had been given colistin was 40 per cent. For plazomicin it was just 14 per cent. That sounded like good news – and it was. But unfortunately, the numbers of patients were tiny: those percentages translated to eight deaths for colistin versus two for plazomicin. These results had cost about $37 million to obtain, but the numbers were so small that they didn't show that plazomicin was better than colistin based on usual statistical thresholds. And that was what Achaogen's trial was designed to test.

It was a harsh judgement, but by the criteria of the trial it was definitive: plazomicin had failed. Achaogen tried to shift the goalposts, arguing that although it had hoped to show that plazomicin would be better, it was reasonable to interpret that it was at least no worse. The weaker conclusion of 'non-inferiority' *was* supported by statistics. Achaogen was appealing to the committee to ignore its original intentions, and to consider the need for an adequate alternative when colistin failed.

Like barristers, the FDA officials cross-examined Achaogen's results. For example, they noted that there was no standardised procedure for distinguishing between different infections in the trial. Even though they recognised that gathering detailed information from intensive care patients might have been an impossible task, it was their job to point out such issues.

The data for the other trial on urinary tract infections was much better, and it seemed highly likely that the committee would vote in favour of that indication for plazomicin. But Achaogen had dreamt of winning two indications at the same time. Its hopes for winning the bloodstream indication, despite all the problems with the trial, hinged on the new regulatory pathway enshrined in law by the Cures Act. But the committee had never before considered a drug under this brand-new legislation and therefore had no precedent to follow. As the committee members debated among themselves, it became obvious that nobody was quite sure what they were allowed to do. At times the debate became almost philosophical. It strained at the whole purpose of a clinical trial. What should the appropriate statistical yardstick be for a new antibiotic that was only meant to be used in situations of life and death?

A non-voting representative from the pharmaceutical industry who was present at the meeting spoke up in favour of Achaogen. It was wishful thinking, he said, to expect a company to provide a comprehensive study of hundreds of patients with an extraordinarily rare and life-threatening infection – other companies had tried and failed. Achaogen had persevered to try and meet an unmet medical need. Was it to be punished for its foresight in developing a new antibiotic, simply because it couldn't be tested as robustly as other medicines? That question was on everyone's mind as the committee broke for lunch. They would return in the afternoon.

After lunch, there was an open public hearing. A scientist called Shoshana Shendelman told a story about her cousin, Fari, who had been a neurosurgeon. After successful chemotherapy for leukaemia, Fari had contracted sepsis because of a CRE infection. His infection was also resistant to colistin and there was apparently no hope. The doctors told Fari's family to say their goodbyes, but Shendelman was aware that new antibiotics were in development. She wondered if they could help him. Quickly, she found out about a new drug – plazomicin – and called up Achaogen to try to access it.

It was a Friday afternoon. The supplies of the drug were locked up in a warehouse in Pennsylvania; doctors told her that if plazomicin treatment wasn't started within twenty-four hours, Fari would die. Late into the night, they'd made urgent calls to the FDA and the hospital board, pleading for permission to give him an unapproved medicine as a last-ditch effort. After receiving plazomicin, her cousin woke up. 'It's important to have as many tools in our arsenal as possible when treating complicated infections,' Shendelman concluded. 'When a patient is a loved one, you start to think about drug development differently.'

But this anecdote, however moving, was not a statistical analysis. Another speaker, a representative from the National Center for Health Research, which had opposed the Cures Act on the grounds that it would weaken FDA standards, told the committee that approving plazomicin with such limited data would set a dangerous precedent. 'We cannot lower the standards,' she said. 'That is what the law requires'. Approving the indication, she was suggesting, would make a mockery of the entire process.

The plazomicin conundrum was exactly the situation that the Cures Act had sought to address, with its provision for a more permissive regulatory pathway for new antibiotics against life-threatening infections. Now, for the first time, that idea was being put to the test. But many committee members were still unsure what was allowed. As the final votes approached, they sought clarification, some of them seemingly hoping to be told what to do.

The elephant in the room was that Achaogen was seeking two indications, and one of them looked certain to pass; nobody was in any doubt that the day would end with plazomicin being recommended for approval with an indication to treat complicated urinary tract infections. And once that happened, there would be nothing to stop physicians using it off-label, as had happened for other drugs in the past, and prescribing it for the very situation they were currently debating. As one of the committee members put it, 'An approval is an approval is an approval'. In this way of thinking, voting against the indication for bloodstream infections wouldn't prevent anybody who really needed it from getting it. Would it?

The time for voting came. First, the committee voted for the urinary infection indication. It passed unanimously: fifteen votes in favour. Plazomicin would be recommended for approval. The second vote came, and the phrasing of the question was crucial: had Achaogen provided *substantial* evidence of the safety and effectiveness of plazomicin to treat bloodstream infections in patients with limited or no treatment options? Four committee members thought yes. But as they went around the room, eleven others voiced their dissent. Achaogen had failed.

The committee members who voted against plazomicin had, minutes earlier, voted for its approval. Some were reluctant, almost apologetic. They had come to the meeting leaning towards yes, but after hours discussing the limited evidence, they felt they had to vote against.

As the news rippled out from the meeting, the world learned that plazomicin had been recommended for approval. Achaogen executives talked about how happy they were that their antibiotic's value was now recognised. Soon, doctors would be able to prescribe it. It should have been a time for celebration.

Instead, it was a disaster: Achaogen shares slumped by 28 per cent. Over the following days, the share price continued to collapse. The committee's 'yes and no' decision was like a red flag to investors that Achaogen's rise was not unstoppable. The more investors looked at the fundamentals of its business model, the more it was clear that something wasn't right.

On paper, the good news kept coming. In June 2018, the FDA followed the committee's recommendation – as expected – and approved plazomicin for urinary tract infections. In August, the US healthcare insurance programme Medicare announced it would reimburse half the cost of plazomicin for every patient – a recognition of its value. But now investors were scrutinising plazomicin's earning potential carefully. Medicare's reimbursement amounted to $3,000 per patient, which was pitiful compared to other potentially life-saving drugs; for some cancer treatments, Medicare would reimburse over $180,000. But if plazomicin was too cheap, it was at the same time too expensive compared to other antibiotics – for example, colistin treatment cost $56 per day – making it unlikely that many doctors would choose to use it. Plazomicin was extraordinarily unlikely to make any money in the foreseeable future.

Publicly, Achaogen executives put on a brave face. But behind the scenes, the company's fortunes continued to worsen. Achaogen was trying to build a nationwide sales force from scratch to compete with big pharmaceutical companies, not to mention setting up global supply chains for manufacturing plazomicin in large quantities. They were having to spend millions – including on standard follow-up studies mandated by the FDA for newly approved drugs – with almost no new investment coming into the company. That was creating a cashflow crisis, which, in turn, scared new investors away. It was a vicious cycle. The company became a perfect target for short selling: traders aiming to profit from Achaogen's situation by betting that shares would continue to fall. The shares kept falling – and the more investors sold them off, the more they fell. Eventually, there was no money left.

Nine months after plazomicin was approved, the downward spiral ended when Achaogen filed for bankruptcy in April 2019. After fifteen years and a billion dollars in funding, plazomicin hadn't even made $1 million in sales. Achaogen shares had once been valued at $25, but they now stood at about two cents. Ryan Cirz, one of the company's co-founders, had worked there from the beginning to the bitter end, and his life savings were in Achaogen shares. He sat in a courtroom for the bankruptcy proceedings, watching nervously as Achaogen's assets were auctioned off, hoping for at least a reasonable buyout that would cover the redundancy package in his contract. He watched in disbelief as Achaogen sold its remaining laboratory equipment, with the global rights to plazomicin thrown in, for $16 million – less than the value of the first grant they'd won from the Department of Defense.

Achaogen had successfully anticipated an unmet medical need, combining cutting-edge chemistry and knowledge of antibiotic resistance mechanisms to modify a molecule evolved by bacteria. Scientifically, it had been a success. But economically, it had been a disaster. As one investor who had invested in Achaogen put it afterwards, the problem for antibiotics was 'screwy commercialization dynamics'. It seemed so obvious in retrospect: if you made a new antibiotic to help a small group of patients, you would first face a struggle to get it approved – and then, if you did, you wouldn't make money. In a market desperate for profit, antibiotics were an appalling investment.

Our world is shaped by the ebb and flow of capital, from the layout of our cities to the medicines in our hospitals. The flow is never wholly rational. In 2001, the anthrax letters killed five Americans; every year, *E. coli* infections kill 36,000. But it was the threat of more anthrax attacks – real or imagined – that had led to the US government allocating billions to Project Bioshield, in the febrile atmosphere of the War on Terror. Achaogen was among the first to work out how to access this money, spinning plazomicin as a matter of national security, allowing it to go beyond venture capital alone and access millions of dollars through government officials. BARDA, the funding agency founded under the auspices of Bioshield, chose to back Achaogen, helping plazomicin to become the first BARDA-funded antibiotic to achieve the milestone of FDA approval. Over the following years, BARDA would invest more public funds in products to combat antibiotic resistance, by 2024 amounting to $2 billion in antibiotics, vaccines and diagnostic tests: over 130 different products. Whether or not American citizens knew it, public funds had been used to prop up companies which, in a truly free market, would have ceased to exist.

Achaogen had once been celebrated as an incredible success story: a billion-dollar antibiotic company. Now, its failure became a cautionary tale, casting an ominous cloud over the future of antibiotics. The hope had been that agile startups would replace the big pharmaceutical companies that had abandoned the arena. But as an editorial in the *Financial Times* put it, Achaogen's bankruptcy showed that was an illusion: however valuable a new antibiotic was to society, there was 'no viable path' for it to succeed. More bad news followed. At the beginning of 2019, three other startups had recently won FDA approval for new antibiotics. Before the end of the year, two had filed for bankruptcy; the third, once valued at $1.8 billion, was bought at a knockdown price. Antibiotics had become a toxic asset. In Achaogen's case, the company that had bought the global rights to plazomicin announced that it wouldn't even seek approval from the European Medicines Agency, the European equivalent of the FDA, because to do so was financially unviable.

In the golden age of antibiotics, capital had flowed into research and development of new drugs, multiplying itself many times over to return astounding profits. The catastrophic end of Achaogen, a glamorous West Coast startup with a patent on a life-saving new antibiotic, was a symbolic moment. In a South San Francisco filled with companies focusing on artificial intelligence and genetic technology, the glory days of antibiotic research seemed as distant as the meat-packing factories and abandoned shipyards that had once been the economic heart of the industrial city.

A few months after Achaogen filed for bankruptcy, its co-founder Ryan Cirz was having breakfast in a cafe when he stared down from his laptop at his receipt. Cirz realised that he'd just paid more for his coffee and mini Danish ($8.84) than for a ten-day course of amoxicillin ($8.39). That was the market that a new antibiotic had to compete in.

The crisis had been described as a market failure, but that was sugarcoating it. There was barely a market for new antibiotics at all.

Realms of gold

Teixobactin

The Lord hath created medicines out of the earth;
and he that is wise will not abhor them.

Ecclesiasticus 38:4

The demise of the market for antibiotics seemed almost complete. And perhaps its downfall wasn't simply a question of capital: antibiotic discovery itself had become more difficult. After the glut of new antibiotic classes from soil in the golden age, the earth appeared to have been mined to exhaustion. Conventional wisdom in the pharmaceutical industry had it that soil screening was an expensive, fruitless form of historical re-enactment. The basics had remained largely unchanged since the golden age of antibiotics – dilute a soil sample, place tiny fractions onto an agar plate to see what kills other microbes and try to isolate the molecules responsible. Compared to testing synthetic chemicals directly, the process was labour-intensive and maddening. The easiest things to find were known antibiotics, made by prolific antibiotic-producing bacteria such as *Streptomyces*. According to one calculation, *Streptomyces* had been responsible for 55 per cent of all the antibiotics discovered between 1945 and 1978. The law of diminishing returns seemed to suggest that soil was over. But in the 2000s, some scientists dared to disagree with this pessimistic assessment. Like medieval navigators looking at the blank spaces on their map, they saw the possibility of new continents as yet uncharted.

The earth, they believed, was far from empty. Only a fraction of

197

bacterial species in water or soil could be grown on an agar plate – perhaps less than 1 per cent. The term for this discrepancy was the Great Plate Count Anomaly. The term dated to the 1980s, but Winogradsky, the grandfather of soil microbiology, had noticed the phenomenon in the late nineteenth century. Yet still the 'unculturable' bacteria had remained largely a mystery. How could you study what you couldn't see? But at the end of the twentieth century, the availability of DNA sequencing allowed microbiologists to analyse the genomes present in soil samples directly. All the earth's organisms left such faint traces behind. Suddenly, the silent majority sang out from the darkness.

The scale of bacterial diversity revealed by DNA sequencing was dizzying. A phylum is a broad taxonomic division; examples in the animal kingdom would be vertebrates or molluscs. Researchers estimated from DNA sequences alone that there were at least fifty bacterial phyla, of which half had never been cultured in a laboratory. The isolated islands of diversity that had been charted by microbiologists were not the whole microbial world. New continents were resolving on the map – not only bacteria, but previously unknown forms of fungi and archaea. A microbiologist could now be in the strange situation of knowing the entire genome of a species without ever encountering it down the microscope, like a biographer who never meets their subject. Yet without the ability to grow a microbe in the laboratory, investigating whether it made new antibiotics was impossible. The genome sequences of these bacteria were proof of their existence, but a poor substitute for studying their cells.

There was a clear precedent for the problem in the origins of microbiology. Obtaining a pure culture of a single bacterial species had seemed impossible before the middle of the nineteenth century: it was only once dedicated scientists like Robert Koch worked out how to grow them that individuals could be summoned out of the multitude. Solving the Great Plate Count Anomaly involved the same basic task. If the 'unculturable' soil species could be grown, they could be studied. And if they could be studied, then their capacity to make new antibiotics could be investigated.

Up until this point, the search for new antibiotics had been like trying to map the earth at night. The dawn might be about to break, flooding the landscape with sunshine.

In 2002, Kim Lewis and his colleagues at Northeastern University in Boston turned microbiology upside down to address the Great Plate Count Anomaly: rather than bringing soil to the lab, they brought the lab to the soil. Instead of thrusting bacteria out on an agar plate, they proposed to take a single bacterial cell from a soil sample and place it inside a small plastic container. The container had a barrier that prevented other bacteria from getting inside, while allowing the flow of nutrients in and out. Then, the container was placed back into the soil itself. The effect of this was to return the cell to its natural environment in a walled garden, where it could be grown in isolation while still taking part in the ebb and flow of nutrients. The beauty of this approach was that it wasn't necessary to know which nutrients the bacterium needed to grow: the soil itself would pro- vide them as it always had. Over time, the bacterium in the container would grow from a single cell to many, producing a pure population that could then be studied. Compared to the 1 per cent success rate of a standard agar plate, Lewis and his colleagues argued they could recover up to 50 per cent of soil's bacterial diversity.

Over time, they refined and scaled up the technology to use the container approach for hundreds of cells simultaneously: a grid of them in a piece of prefabricated plastic the length of a finger which they called the isolation chip, or iChip for short. A decade later in 2015, they described how they'd collected a pinch of soil from a grassy field in Maine – in fact, from the lead author Losee Ling's backyard – and then diluted it in water to create a faint suspen- sion of bacterial cells with one bacterium every twenty microlitres. Each hole in the iChip got a twenty-microlitre droplet, containing on average a single cell, before the bacteria were sealed into their separate prisons. Then, the iChip was placed back into the soil for a month to incubate these separate and unknown species while the researchers waited.

From that initial pinch of soil, they obtained 10,000 different bacterial isolates – a large number, but a manageable one. They tested the bacteria for their ability to inhibit the growth of *Staphylococcus aureus*, the species that includes MRSA. The most successful species was a previously unknown bacterium, and it made a new antibiotic. Searching for a name for their discovery at a time of increasing alarm about resistance, the researchers named the molecule after the Greek *teixos*, the fortified wall protecting a city: they called it teixobactin.

Teixobactin appeared to be something amazing, a 'resistance-proof' antibiotic. The researchers had exposed staphylococci to teixobactin for a month and failed to find any detectable resistance. Lewis, the senior author on the paper, suggested that teixobactin challenged the dogma that bacteria would always evolve resistance. He was (predictably) wrong – subsequent studies showed that staphylococci could in fact evolve teixobactin resistance. However, it happened hundreds or thousands of times more slowly than for some other antibiotics. Lewis and his co-authors hadn't recovered resistance in their own experiments because, on that timescale, the cloud of genetic variation generated wasn't large enough to encompass a teixobactin-resistant bacterium. As always, evolution still applied: a resistance-proof antibiotic was like a free lunch in economics. Time and time again, bacteria had shown that no antibiotic was immortal. However, a low rate of resistance was much better than a high rate, and it suggested teixobactin might enjoy exceptional longevity as a medicine.

As a new antibiotic, teixobactin was a proof of principle that the soil had not been mined to exhaustion, but it was also a perfect example of the excitement and frustration that characterised a natural antibiotic. Isolating the single molecule responsible for the effect from the soup of chemicals present in the experiment, and then working out its structure, had been a major undertaking. Methods were improved from the days when Dorothy Crowfoot Hodgkin had needed to run computations on a punch-card machine to confirm penicillin's structure, but the process was still not trivial. When it was obtained, teixobactin's structure had all the hallmarks of a natural molecule.

A human would never have designed such an ungainly construction, complete with a giant ring and unusual molecular components. One medicinal chemist working in pharmaceutical development, Derek Lowe, called the structure 'pretty chewy'.

Presumably, just like the beta-lactam ring at the heart of penicillin, teixobactin's confusing structure was crucial for its effectiveness. But again, it looked challenging to synthesise chemically at a reasonable cost. None of these problems were conceptually new in antibiotic development. In principle, they could be solved for teixobactin just as they had once been solved for penicillin. But these scientific challenges were only one aspect of the problem. If teixobactin was going to be a medicine, how would it get there without investment? As startups like Achaogen visibly struggled to survive, the broken development pipeline for new antibiotics was in desperate need of reform.

When teixobactin was discovered, Kevin Outterson was working as a law professor at Boston University. As an expert in intellectual property, he'd had his attention drawn to antibiotics by the case of vancomycin. The American company Eli Lilly had obtained the antibiotic from soil sent from the jungles of Borneo by a missionary in the 1950s. Vancomycin had been patented at the time but rarely used. Only decades later, when the need arose to treat antibiotic-resistant bacteria like MRSA, did it become most valuable to society – years after its patent had expired. That story had been repeated for other antibiotics, such as colistin. If antibiotics were innovations whose social value might lie decades in the future rather than in the present, then patents were an ineffective instrument: an antibiotic was not a new type of dishwasher, but a long-term public good. Outterson saw that despite the universal agreement that new antibiotics were needed, small companies trying to make new antibiotics regularly went bankrupt at the early stages of development. All too often, they simply ran out of money.

To help address that problem, Outterson launched a non-profit partnership called CARB-X in 2016. CARB-X had a simple premise:

it would fund companies with research grants when other private investors wouldn't. The money for CARB-X came from the US, UK and German governments as well as three charitable foundations: Wellcome and the Novo Nordisk Foundation, both originally founded with pharmaceutical profits, and the Gates Foundation. The resulting model of funding was a strange mixture of public and private, with the same dollar sometimes changing hands (and acronyms) several times before it was used directly on antibiotic research. Money from the American public could go to an agency originally tasked with protecting citizens against bioterrorism (BARDA), before being given to a non-profit organisation (CARB-X), which would then award it to a private company.

CARB-X was not the only attempt to boost antibiotic development. Pharmaceutical research by companies delivers drugs based on what is profitable, rather than what is most urgently needed, and the problem had long been especially acute for the needs of people in poorer countries far from the global centres of pharmaceutical power. When Médecins Sans Frontières (MSF) won the Nobel Peace Prize in 1999, the president of their International Council had used the opportunity to call attention to the fact that the majority of death and suffering from infectious disease happens in poorer countries, driven by both a lack of access to existing medicines and a lack of research into new ones. 'This market failure', he told his audience, 'is our next challenge'. MSF used the prize money to help set up the Drugs for Neglected Diseases initiative in 2003. In 2016, DNDi worked alongside the WHO to set up an organisation applying the same model to antibiotics.

The new organisation was called the Global Antibiotic Research and Development Partnership (GARDP). Where CARB-X focused on the early stages of antibiotic development, GARDP would focus on the later ones, where expensive clinical trials were needed. The aim was to help promising antibiotics win approval by reducing the risk of the trials, partnering with companies by providing funding and technical assistance in return for the companies giving up their rights to market access in poorer countries – which companies weren't usually interested in anyway. GARDP would then own those rights and be able to sublicense them to other manufacturers.

CARB-X and GARDP represented efforts at different ends of the antibiotic development pipeline, but both were driven by public organisations trying to inject more money into the pharmaceutical industry. The pharmaceutical industry was more reluctant to invest further money of its own. In 2019, Jim O'Neill, who had authored an influential review commissioned by the British government into antibiotic resistance, pronounced himself unimpressed with the progress in the private sector. 'Nearly three years after our review came out, there's endless talk but there's no progress,' he told the BBC. To increase the number of new antibiotics, his review had suggested a global innovation fund for antibiotic research and better incentives for investment. But now he made a startling suggestion: he felt it might be time to 'just take it away from them and take it over'. O'Neill, an economist who had spent his career at Goldman Sachs and served as a minister in a right-wing Conservative government, was prepared to countenance the idea of the state developing antibiotics itself rather than propping up private companies with public funds. He admitted that he would once have found the idea 'a bit crazy'.

The industry was anxious to prevent O'Neill's idea becoming more popular. In 2020, a group of over twenty pharmaceutical companies announced the formation of the AMR Action Fund, with an investment pool of nearly $1 billion specifically for research on antibiotic resistance. In 2021, a further $140 million was added by the European Investment Bank and two charitable foundations. The fund's stated purpose was to invest strategically in order to get between two and four new antibiotics to patients by 2030, while still making a return for its investors. That made it a substantially different proposition to CARB-X and GARDP. The fact that companies were cooperating indicated that this may have been more of a collective PR effort in damage control than a truly transformative investment.

The AMR Action Fund fulfilled an important strategic need for large pharmaceutical companies: when challenged about their lack of investment in antibiotics, they now had something concrete to point to. The size of each company's investment wasn't made public, but splitting the fund's initial value of $1 billion between more than twenty companies meant that the average investment had to be less than $50 million. In the context of drug development – which the industry claimed to cost an average of over $2 billion per drug – that was not a large amount. The AMR Action Fund was evidence that pharmaceutical companies would not, on their own, try to solve the structural challenges that were thwarting antibiotic development.

Meanwhile, scientists were still struggling to conceptualise antibiotics. The way teixobactin worked to attack lipids in the bacterial cell revealed new aspects of biochemistry – but even that framing was only a partial truth. All the way back to Ehrlich's magic bullet metaphor, scientists had understood antibiotics as weapons with specific targets. That was what had influenced the idea of first identifying targets, then finding molecules that attacked them. But that rational way to approach the problem of finding new antibiotics had totally failed. The failing was an all-too-human one. As thinking beings, humans approach the problem as designers, but antibiotics were not *designed* by evolution in this way.

The whole metaphor of a magic bullet with just one target may well be mistaken. The drug developer Lynn Silver has argued that effective antibiotics rarely have just one target. An antibiotic can have multiple effects on a bacterial cell – what biologists call 'pleiotropy', from the Greek *pleion* ('more') and *tropos* ('turn'). A good antibiotic will be a molecule with many facets, like a crystal that catches light from different angles. An antibiotic with just one target is vulnerable to resistance; often a single mutation in the bacteria's genome may render it ineffective. An antibiotic that is pleiotropic can't be evaded so easily. It's as if a pleiotropic antibiotic can act as a combination therapy despite being only one substance. Unlike a human designer, evolution has no guiding design principles. It does not draw up a molecule to fit a single pre-specified function; the molecules it produces endure because the organisms that make them survive. That means that natural antibiotics are much more likely to have this pleiotropic property than molecules designed by humans.

The difference highlights the absurdity of claiming mastery over bacteria. Despite the founding of new organisations, society has not yet reckoned with the true difficulty of developing new antibiotics. CARB-X's founder Outterson has estimated that obtaining six high-impact antibiotics per decade – ones that truly made a difference to patients – would require undertaking six *thousand* basic research projects. Of those, about 200 per decade would progress to early-stage development, with around twelve making it to clinical trials. In 2015, teixobactin had been heralded as the founding member of a new class of antibiotics. The startup behind it, NovoBiotic, had received millions of dollars in funding from the US government including for research using teixobactin against anthrax. But progress was slow: in April 2023, just after announcing a further $4.5 million in funding, teixobactin had still not started even Phase 1 clinical trials. An antibiotic that had stirred hope for the future almost a decade before was yet to make it into a human body.

Compared to plazomicin and other antibiotics which were modifications of known antibiotics, teixobactin's very novelty was a blessing and a curse. Its unfamiliar structure raised more questions than answers. Lowe, the medicinal chemist who called its structure 'chewy', said that while he wished NovoBiotic well, decades of experience showed that most large molecules like teixobactin had 'awful stability and pharmacokinetics', making them challenging to develop into useful drugs.

At the time of writing, it is impossible to tell whether teixobactin could be among the lucky molecules that make it all the way to market. It wouldn't transform medicine even it did, though it would likely provide an alternative and welcome treatment for MRSA compared to vancomycin. Despite the huge fanfare accompanying the discovery of a new antibiotic, such a discovery is never life-saving on its own but only a tentative early step on the long journey to a new medicine. Researchers are desperate to improve the chances of getting from that initial molecule to an approved medicine, but it's clear that current scientific understanding of the complexity involved is pitiful. In 2024, researchers working with GARDP tested over 48,000 molecules against drug-resistant Gram-negative bacteria like CRE. The molecules were provided by three pharmaceutical companies from their chemical libraries, deliberately chosen because they were believed to be more likely to get inside the bacterial cells. The approach totally failed: not a single molecule out of 48,000 was promising enough to take further into preclinical development.

When the chemist Heinz Moser worked at Achaogen before it went bankrupt, he had published an analysis of existing antibiotics to try to better understand why developing new ones was hard. Of the twenty antibiotic classes he analysed, just three had been discovered by chemistry rather than from nature. Comparing the molecular structure of antibiotics to other medicinal drugs, it was clear that they were very different. Chemists had previously tried to codify rules to guide future discovery. The most famous was known as Lipinski's rule of five, formulated in the early 2000s, comprising five properties that molecules tend to have if they are effective oral drugs. Moser's publication made it clear that antibiotics violated Lipinski's rule of five. The only antibiotics that obeyed the rule were those from the oldest class: Prontosil and its other sulfa relatives. Put simply, most antibiotics didn't look like drugs. The most effective among them had been invented not through design but by the trial and error of evolution, and then discovered by humans in much the same way. As the drug developer Lynn Silver had pointed out, in contrast to modern attempts at 'rational' drug design, the older and 'irrational' approach of testing molecules in the laboratory had at least one thing going for it: it had worked.

But now, every road to a new antibiotic presented drug developers with difficulties. Teixobactin showed that the soil was not exhausted, but its new and unfamiliar molecular structure meant turning it into a medicine was inherently more uncertain. Variants of known antibiotics like plazomicin or new beta-lactams were more reliable – but consequently, they offered less transformative potential for medicine and were prone to similar forms of resistance. And the most outlandish approach, designing new antibiotics altogether, appeared impossible, because what made a successful antibiotic remained mysterious.

Some scientists dared to ask an almost sacrilegious question: was it possible that antibiotics themselves were outmoded? Perhaps it was time to stop chasing the dream of new antibiotics and move on to other technologies. After all, Ehrlich's concept of a magic bullet was over a century old. Medicine's desperate reliance on a concept borrowed from an old folk tale was beginning to look antiquated. The most serious cases of multi-resistant infections didn't need a new treatment that would become available in over a decade – they needed one now. Rather than relying on a slow conveyor belt of new antibiotics, what if doctors could generate a bespoke treatment for a patient's infection in days rather than decades?

When Steffanie Strathdee's husband Tom fell ill on a Nile cruise in 2015, it had started mildly enough with stomach cramps. But his condition worsened, and he was flown to Germany for medical treatment. Doctors there discovered an abdominal abscess filled with multi-resistant bacteria, and he was flown again to San Diego. Tom's infection was caused by a bacterium called *Acinetobacter baumannii*. In the Vietnam War, *Acinetobacter* was the most common bacterium in traumatic injuries to extremities. But after the 2003 invasion of Iraq, drug-resistant strains were observed infecting the wounds of American soldiers, and the bacterium became known as 'Iraqibacter'. (There was an irony that a war meant to stop a regime with supposed bioweapons had ended up exposing American soldiers to a natural drug-resistant threat.)

Wars have a track record of generating new drug-resistant threats by combining extreme injuries, insanitary conditions and crumbling health systems. Those threats spread: drug-resistant *Acinetobacter* was now a growing scourge of intensive care units around the world with mortality rates of up to 35 per cent. The doctors at

UC San Diego Medical Center were at the forefront of US health-care. But months of antibiotic treatment didn't work, and Tom's organs started to fail. In February 2016, doctors told Strathdee that her exhausted husband, unable to speak, was going to die. She held his hand and told him, 'If you can hear me and you want to live, please squeeze my hand.' He squeezed back.

Strathdee felt she had a responsibility to help Tom live. But without new antibiotics, what could she do? She had an advantage: she was an academic in health sciences and an associate dean at UC San Diego. She quickly decided that the only option was to try using viruses to attack the *Acinetobacter*.

Bacteria have their own viruses, referred to as bacteriophages, or just 'phages', because they were originally thought to consume bacteria (the Greek *phagein* meaning 'to eat'). Phages inject themselves into the bacterial cell where they begin replicating using the bacteria's own cellular machinery, before bursting out and infecting more cells: a microscale viral epidemic.

A century before, phages had been the future. In 1925, the novel *Arrowsmith* by Sinclair Lewis had been an instant bestseller. The book told the life story of a fictional scientist, Martin Arrowsmith. As a young doctor, Arrowsmith is distraught by medicine's failure to treat infectious diseases. He tries to save a young girl from diphtheria, driving across the countryside in the middle of the night to fetch the precious supply of antitoxin. But he arrives too late, and the girl dies. Motivated by his failure, Arrowsmith makes a sensational discovery: he discovers bacteriophages and finds they can vanquish infection. After plague breaks out on a Caribbean island, Arrowsmith rushes there to trial his new therapy. Moved by the suffering he sees, he goes against his scientific training and abandons the controlled trial, treating everyone he can to save as many lives as possible.

Lewis's novel got it wrong – antibiotics, not phages, would transform medicine. The discovery of Prontosil in the 1930s followed by the mass production of penicillin meant that, in most parts of the world, phage therapy became outmoded. Over the latter half of the twentieth century, phages didn't receive the medical attention they had once warranted. It was only after the rise of antibiotic resistance and cases like Strathdee's husband Tom that phages offered a slender thread of hope.

Compared to antibiotics, phages are highly specific in the bacteria they target. To save her husband Strathdee needed to find a phage that could kill the exact strain of *Acinetobacter* that was killing him. In fact, she needed to find several. A bacterium can quickly become resistant to a phage by discarding the surface molecule the phage needs to latch on to it. Using the same principle as combination therapy for antibiotics, avoiding resistance requires multiple phages. Strathdee phoned and emailed phage scientists across the world and asked them for their help: she enlisted scientists from the US Navy and Texas A&M to produce 'cocktails' of phage that would attack Tom's infection. But because those new cocktails would count as an untested therapy, even though Tom was dying she needed to deal with a mass of legal paperwork before doctors could administer it. There she had another advantage: she 'literally had speed dial numbers for the chancellor and all the people involved in human experimentation'. As Tom lay in a coma, scientists prepared two different phage cocktails. Three days after the cocktails were injected into his body, he woke up and kissed his daughter's hand. Over the next few weeks his condition continued to improve. Thanks to Strathdee's efforts he had survived an infection that would otherwise have killed him.

Strathdee had managed to create something almost impossible: a new bespoke therapy that had been immediately injected into a patient. She believed her husband was the first person in the US who had received intravenous phage therapy for a systemic infection with a superbug. A scientific paper published about the results in 2017 awoke new interest in phages. But without a wife who was a powerful academic, Tom would have died. The creation of the phage cocktails relied on groups of scientists working around the clock for a single patient; the morass of medical paperwork had been dealt with at the highest possible level. Tom's survival was a remarkable triumph, but it could not serve as a blueprint for routine medical care.

However, early uses of antibiotics had also been beset by similar problems. And after Tom's treatment, phage therapy saw a renaissance. In a review of the newly accelerating field, Strathdee and her co-authors wrote of their optimism that the limitations that had thwarted phage therapy would be overcome with new technological advances. Phages presented exciting possibilities beyond simply treating severe multi-resistant infections. Some phages use a surface molecule linked to the transfer of plasmids between cells, meaning that they could potentially target plasmids carrying antibiotic resistance genes. Or they could be used in gels to avoid infections on prosthetic joint surfaces, to disinfect wastewater without chemicals or to precisely manipulate the composition of the gut microbiome.

The number of new clinical trials involving phage therapy grew. But as leading phage scientist Martha Clokie acknowledged, antibiotics were far easier to use as medicines. The world still needed new ones. In 2020, one group of scientists felt it was not yet time to give up on finding magic bullets. They hoped to rejuvenate antibiotic discovery with a dramatic claim: that they had discovered a new antibiotic, not from nature, nor by human design, but by using artificial intelligence.

One wide expanse
Halicin

One should sit quietly and let the thing invent itself.

Iris Murdoch

Chemistry, like politics, is the art of the possible. Chemists sometimes talk of 'chemical space': an imaginary zone of possibility inhabited by all the molecules that could ever exist. It's difficult to get a grip on how vast this zone is. Most drugs have fewer than thirty atoms. Even with a handful of elements – carbon, hydrogen, oxygen, nitrogen, sulphur – there are at least 10^{60} possible molecules, more than there are atoms in the sun. Most of these hypothetical molecules will never be. Chemical space is like the library imagined by Jorge Luis Borges: a building of unimaginable size that contains every possible book. All the books that have ever been or ever will be written by a human hand are an infinitesimal fraction, and all the generations of humanity could pass in front of its shelves without seeing a single page they recognise.

Rather than books on a shelf, a better metaphor for chemical space might be stars in the night sky – not uniformly arranged, but clumped. There are dark zones of impossibility, interspersed with bright patches where molecules cluster into galaxies around central motifs. (But there are far more possible molecules in chemical space than there are stars in our universe.) Of the molecules that have ever existed on Earth, most have been made by living things. The atoms on Earth have barely changed in 4 billion years, but life has rearranged them again and again into different molecules. The imprint of life even extends into geology: without life, half of all minerals on Earth would not exist. From the basic molecules of life, evolution has explored outwards through the darkness of chemical space, building trading routes between different galaxies. It has generated the diversity of molecular structures found in nature, from the luciferin that glows yellow in fireflies to the green chlorophyll in an unfurling fern.

The antibiotic classes in chemical space form galaxies of proximity thanks to their central motifs – the beta-lactams clustered together in one region, the aminoglycosides in another, and so on. Where chemists take an antibiotic and tweak it to stave off resistance or improve its properties, they are only creating a new star in an existing galaxy. Finding new galaxies is much harder. Searching in nature for antibiotics is valuable precisely because evolution has been exploring the possibilities of chemical space for billions of years. Even synthetic antibiotics like Prontosil or isoniazid originated from organic molecules: hydrocarbons from oil led to Prontosil; isoniazid is a derivative of a substance that can be found in the bitter leaves of the aloe plant. In the words of Margaret Chan, who served as the WHO's Director-General, 'All of the "easy" antibiotics have already been discovered.' The furthest gulfs of the chemical universe could conceivably contain new antibiotics, but if evolution has failed to discover them, a human lifespan seems unlikely to. Finding new antibiotics through a systematic search is as impossible as constructing the total library of Borges: the entire age of the universe wouldn't be enough.

In 1960, a wealthy industrialist in Los Angeles donated a state-of-the-art computer to the chemistry department at the local Pomona College. The Clary DE-60 computer, one of fewer than twenty such devices in existence, cost nearly $200,000 in modern terms. The chemists in the department had little clue what to do with it. The academic who had persuaded the industrialist to donate the DE-60 explained that they could use it to run calculations. The chemists were trying to predict the potency of different molecules – high potency being a desirable feature for a new drug – from their chemical properties alone. Currently, that meant using a mechanical calculator by hand. The process took hours and could be thrown off by a single mistake; on one model, accidentally dividing a number by zero would require phoning a repairman to come and reset the machine.

Their colleague helped them use the DE-60 to run a large analysis that would have been impossible without it. By fitting a statistical relationship to what was then a large amount of data, they discovered a new law of chemistry. It would speed up drug development, allowing them to predict how potent a molecule would be before it was even synthesised.

The Clary DE-60 had been used to analyse just twenty molecules. As the power of computers grew, so did their application throughout chemistry, building on that first analysis and refining methods to predict the properties of unseen molecules. In 2020, researchers at MIT sought to do the same: to predict whether a molecule would work as an antibiotic *before* testing it against bacteria. First, they needed to generate data to train their algorithm, so they tested 2,500 molecules against *E. coli*, one of the species within CRE. Jon Stokes, the researcher tasked with running the main experiments, thought at first it was a tedious project that he'd spend at most a week on.

A molecule with antibiotic activity in the laboratory will not necessarily make a good antibiotic – and most do not. Just as hearing the first chord from the orchestra does not reveal the whole symphony, initial antibiotic activity is only one of the many properties an antibiotic needs. Once a molecule is administered to a patient, the complexity of molecular interactions quickly spirals out beyond human comprehension, and a single problem will doom its use as a medicine. From Prontosil onwards, putting an untested molecule inside a human body has tended to produce unforeseen consequences. The history of antibiotics is littered with failed candidates.

For this reason, the MIT team applied their algorithm to a database of molecules known as the Drug Repurposing Hub. The hub can be thought of as a library detailing the recent history of drug development, containing over 7,000 molecules that have been previously studied as potential drugs. Though not investigated as antibiotics, the hope was that there might be molecules that could serve as antibiotics lurking in there, just as synthetic dyes had once led to Prontosil. As Stephen Fry once wrote: 'An original idea. That can't be too hard. The library must be full of them.'

The algorithm predicted that ninety-nine of the molecules in the Drug Repurposing Hub would have antibiotic activity against *E. coli* if tested in the laboratory. When Stokes and his colleagues checked the validity of this claim by actually running the experiments with these previously untested molecules, about 50 per cent of them did. In the context of drug discovery, those were good odds. However, when it came to choosing between them, which should be followed up with further investigation? The team's motivation was to identify a new type of antibiotic, so they identified the molecule that was the most different in its structure from known antibiotics.

In the universe of chemical space, they had found a lone star. Their chosen molecule was far away from other antibiotics, suggesting it worked in a new way as yet unknown to science. Further tests showed the molecule could inhibit other pathogenic bacteria as well as *E. coli*. It might be the founding member of a new galaxy of antibiotics. The molecule had a laboratory name that was a string of characters, but in a clever piece of PR, the researchers named it halicin after HAL, the computer in Stanley Kubrick's *2001: A Space Odyssey*.

Newspaper headlines dutifully proclaimed that 'artificial intelligence' had been used to discover a new antibiotic. The MIT researchers argued that their way of using AI would speed up drug discovery. They demonstrated that they could extend their computational screening to predict the antibiotic activity of over 107 million molecules – one hundred times the number that could feasibly be tested in the laboratory – in just four days. The researchers set up a non-profit organisation called Phare Bio, announcing their intention to take halicin forward. They also promised to use their approach to find yet more new antibiotics. If they were successful, it could herald a new era of antibiotic discovery.

Phare Bio's stated aim was memorable: they wanted to discover seven new classes of antibiotics against seven different bacteria in seven years. In the typically slow world of drug development, such ambition was designed to raise eyebrows. By 2023, the team announced they had discovered a second antibiotic candidate among the molecules in the Drug Repurposing Hub. This time, they had discovered a molecule that could kill *Acinetobacter baumannii*, the bacterium that had nearly killed Steffanie Strathdee's husband Tom before his treatment with phage therapy. Phare Bio called this second molecule abaucin. Abaucin seemed to target *Acinetobacter* but had no effect on other related Gram-negative bacteria. In this sense, it was more similar to the highly targeted effects of phage therapy than to a typical antibiotic. In an age of antibiotic resistance, that specificity was an advantage; using abaucin would be less likely to produce resistance in other bacteria that could then spread between species on plasmids.

However, the identification of halicin and abaucin was double-edged. Both molecules had been in the Drug Repurposing Hub because they had once been investigated as human drugs; other researchers had chosen them because they hoped they would interact with human cells, not bacterial ones. Since Ehrlich's conception of magic bullets, a principal tenet of antibiotic discovery had been to choose molecules that should have minimal interactions with human biology. Toxic antibiotics like colistin showed the dangers of using those that affected human cells too.

Previous research on halicin illustrated this difficulty. Six years before the MIT team had identified halicin's antibiotic properties, a different group of researchers studying heart disease had predicted from laboratory experiments that it would improve recovery after a heart attack. When they ran animal trials using halicin, this turned out to be wrong. Experiments inducing heart attacks in rats had shown that halicin unexpectedly starved cells of energy, *worsening* the heart's recovery. In the body, halicin had done the precise opposite of what they had predicted.

Not everyone within the world of drug development was impressed with halicin or abaucin. The assessment of the drug developer John Rex was that halicin was 'just another toxic chemical' – something that killed bacteria in the laboratory, but would turn out to be toxic to humans as well. And the more one dug into the details of both abaucin and halicin, crediting AI with the discoveries began to feel inaccurate. Computer models need data. The AI model that had predicted halicin depended on thousands of carefully collected results. The project hadn't fundamentally changed the tactics.

Phare Bio claimed that AI would speed up the early stages of antibiotic development. They hoped to take two years off the initial stage before clinical trials, where molecules typically spend about four and a half years. Yet unfortunately, their own experience with halicin would not bear this out. The original discovery was published in 2020, and yet in 2024, halicin had only progressed to animal studies. Phare Bio also said that progressing an initial discovery to its first safety trials in humans costs around $2.5 million, and they hoped that using AI could achieve that at a third of the cost. Unfortunately, the cost of getting to safety trials is the cheapest part of drug development.

Pharmaceutical companies are tight-lipped about the details of drug development, but they frequently publicise how expensive it is – they have a vested interest to exaggerate, because it justifies high prices for new medicines. At the Tufts Center for the Study of Drug Development in Boston, health economists produce estimates of the cost. Though Tufts is independent, around 55 per cent of the Center's funding comes from the private sector, and it relies on figures provided by drug companies to do its research. Unsurprisingly, its estimates are high – $2.6 billion per drug according to 2014 research. That isn't the figure for each successful drug – it takes into account all the failures as well – but it was heavily criticised by some, including MSF, which commented 'if you believe that, you probably also believe the earth is flat'. Other estimates are as low as $161 million. But even this is still a significant sum. As a non-profit organisation, Phare Bio could expect to face significant future hurdles in financing the later stages of development.

Despite its admirable ambitions, Phare Bio showed no sign of being able to fix the fundamental issues that had frustrated the recent history of antibiotics. The crisis continued. Rather than designing a new antibiotic, some felt the real challenge was far greater: redesigning research. That meant scrutinising the structures, not of chemicals, but of the pharmaceutical industry.

As an analysis by the management consultancy firm Deloitte puts it, 'Health care is big business – really big.' The biggest chunk of that business remains the US. The American healthcare market acts like a black hole, its gravitational pull warping the global industry. In particular, the FDA approvals process influences which drugs become available to the rest of the world – it is effectively a global regulator. More than half of all new drugs since 2018 have been launched first in the US, with an average lag of a year before their launch in other rich countries. The FDA doesn't just set the initial bar that a new drug has to clear but also the many subsequent stages; even once a drug has been approved, manufacturers must conduct further studies if the FDA tells them to. Achaogen's co-founder Ryan Cirz has estimated that a company that wins approval for a new antibiotic must typically spend between $40 million and $170 million on required studies. Such costs are difficult to bear unless the company is extremely large and robust to potential market shocks (as Achaogen had found).

Over time, the world's biggest pharmaceutical companies have become even bigger. Repeated corporate mergers have concentrated power into fewer and fewer hands. Mergers are times of great stress for people who work at the companies involved, and often disruptive for existing research programmes. In 1970, the Swiss companies CIBA and Geigy, both based in the city of Basel, had different cultures that were considered to be as distinctive as different nationalities. When they unexpectedly announced that they were merging, it felt like the end of the world for some employees: people in Basel still tell stories of employees standing at street corners and throwing confidential papers into the air. The new company that resulted, Ciba-Geigy, seemed gigantic at the time. But in 1996, Ciba-Geigy merged with another Swiss company Sandoz to form a new company, Novartis, which became the second-largest pharmaceutical company in the world. At the time it was perhaps the largest merger in corporate history. In 2018, Novartis announced it was ceasing its antibiotic research programmes.

These mergers have diminished the diversity of the pharmaceutical sector. In the US, only a quarter of the companies that were members of the Pharmaceutical Research and Manufacturers of America industry lobby in the late 1980s are still around today. Plotting out company mergers over time creates a diagram that resembles cells blobbing together into superorganisms. Twenty-seven pharmaceutical companies that existed at the beginning of the 1990s had, by the early 2000s, coalesced into the world's eight largest firms. Glaxo-SmithKline arose from a double merger: SmithKline and Beechams merged in 1989, Glaxo and Wellcome merged in 1995, then the two mega-companies merged in 2000. Pfizer, one of the world's biggest, has swallowed four other companies since 1995.

John LaMattina, a former president of research and development at Pfizer, has called mergers 'bad for science, bad for patients, bad for medicine'. Evidence bears this out. In 2019, researchers analysed patent filings to show that, four years after a merger, there was an average decline in patents per year of more than 20 per cent at the merged companies. The company's competitors also saw an innovation decline of 16 per cent, suggesting that when innovation drops at one company, its rivals can afford to focus less on innovation too. These repeated mergers have turned the pharmaceutical sector into what economists call an oligopoly: a market controlled by a few companies. Large companies dominate the global market, and the barriers to entry for new companies are extremely high, because of the sector's heavy reliance on costly research and development. As judged by inflation-adjusted spending, the industry at the start of the 2020s spends ten times as much on research as it did in the

1980s. But what matters is not just the total amount, but where that research is being directed.

The universe of chemical space is not being explored equally. Corporations, bound by law to pursue shareholder value, are machines for converting human ingenuity into profit. The lack of perceived profitability in antibiotics has caused a brain drain from antibiotic research. The number of researchers working on antibiotic resistance worldwide is estimated to be only 3,000, compared to more than 40,000 researching cancer. The results for medical innovation are unsurprising: in 2022, there were twenty times more patents awarded for cancer than for antibiotics. The majority of antibiotic development that happens now takes place outside large pharmaceutical companies. As much as 80 per cent of antibiotic development is being carried out at companies with fewer than 250 employees. The BEAM Alliance is a network of European companies researching new antibiotics; more than half have ten employees or fewer. When surveyed, more than 50 per cent of companies reported they had less than €1 million in the bank. Such a precarious financial position means these companies can easily go bankrupt long before they progress their research, despite the additional funding routes now available thanks to organisations like CARB-X.

When a company working on antibiotics sinks, its employees have to jump ship. Each move leads to attrition, and many researchers end up leaving antibiotics behind. One 2024 analysis suggests that of those working in antibiotic research, only 10 per cent remain in the field after their second job change. With the loss of those researchers goes not only their years of experience, but the possibility of training the next generation of scientists to tackle the problem. When Achaogen developed plazomicin, the company might have been new, but it contained decades of accumulated human experience in drug development. It had George Miller, the chemist near retirement who was old enough to remember the unfashionable chemistry that the industry had forgotten. And it had Heinz Moser, another experienced chemist. When plazomicin was tested for potential toxicity against the kidneys, Moser had known to run preliminary tests in older rats. He had seen other companies running the same toxicity tests in mice or young rats, which sounded similar enough but to any seasoned drug developer was 'completely stupid': their kidneys regenerated too quickly and a drug toxic to humans would not be picked up. With experts in antibiotic development retiring or moving into other areas, it was possible that AI-aided antibiotic discovery would simply make it easier for inexperienced scientists to set off eagerly down roads that led nowhere.

The crisis in antibiotic development raised an almost philosophical question: what are antibiotics? It was hard to imagine a more perfect anti-capitalist commodity. A new antibiotic would be taken by a patient usually for only a few days, rather than for months or years like a drug for a chronic condition such as diabetes. It would be competing against decades-old antibiotics that were extraordinarily cheap. And worst of all, it would probably be used as little as possible to safeguard against future resistance – especially if it was genuinely innovative. Yet for all that, it would need up-front investment like any other sort of drug. That equation did not balance. Antibiotics did not resemble other drugs at either a molecular or an economic level; a functioning pharmaceutical market for them did not exist. In a meaningful sense, antibiotics were not drugs.

People searched for different metaphors to gesture towards their unique role. In the favoured description of the drug developer John Rex, antibiotics are the fire extinguishers of medicine. Imagine if the manufacturers of fire extinguishers only made money when one of their extinguishers was used to put out a fire. That would be absurd. But that is the situation for antibiotics: pharmaceutical companies only profit if they are prescribed. Why should that be true? Like fire extinguishers, the value of antibiotics lies in their availability; like fire extinguishers, using them rarely is a good thing. What would a business model similar to fire extinguishers – based on provision rather than use – look like for antibiotics?

One possibility would be subscriptions. In such a system, a company that developed a useful antibiotic would be paid a fixed amount each year, rather than for each dose of the antibiotic used. In 2022, the British government began to trial a pilot subscription scheme for two antibiotics. Shionogi (which makes cefiderocol) and Pfizer (which makes ceftazidime–avibactam) both signed a contract with NHS England to be paid £10 million each per year for a decade. The amount was calculated from a hypothetical global subscription model produced by a group of economists, with each country paying a share proportionate to its population. England's socialised healthcare system made such mass purchasing of drugs straightforward. For a year of good life, the National Institute for Health and Care Excellence (NICE) is willing to pay up to £20,000. NICE estimated that paying £10 million per year for each of the two antibiotics, based on expected use within the healthcare system, represented value for money.

The subscription model is a serious attempt to reckon with the unique features of antibiotics. The scheme has been hailed by many as a success, but it's too early to judge whether it's persuading other pharmaceutical companies to invest in developing new antibiotics. Through the scheme, £10 million a year will be paid to two of the world's largest pharmaceutical companies for antibiotics that are already available. That is intended to coax other companies back into antibiotics, but it's not obvious that will happen.

An alternative model would be a one-off prize for a valuable new antibiotic. That prize could take a number of forms. One option would be a direct cash payment. The British government has already tried a prize scheme for a technology to address antibiotic resistance, announced in 2014. After a decade, it awarded the £8 million prize to a company that had developed a faster diagnostic test for bacterial infections, which would increase doctors' confidence when selecting antibiotics to prescribe. A one-off prize for a new antibiotic would need to be substantially bigger than £8 million to incentivise the world's biggest pharmaceutical companies. But giving companies public money as a prize, rather than as payment for a service, is understandably an unpopular policy. The alternative to a direct cash transfer is for governments to stealthily provide financial benefit through other means, such as by tweaking the patent system.

As Outterson, founder of CARB-X, had argued, patents were not working for antibiotics but were extremely lucrative for other new drugs. What if gaining approval for a new antibiotic was rewarded with the prize of a voucher that could be used to extend the patent rights on *another* drug? The voucher would simply be an intangible award of intellectual property, which wouldn't cost a government anything up-front but could be worth billions of dollars – if, for example, the company applied that voucher to extend their exclusivity on a drug for diabetes. Under some variants of this proposal, the company would be allowed to auction off the voucher to the highest bidder. It was an ingenious suggestion that would cost a government nothing to award – at first.

But the value of the voucher was that it would make healthcare more expensive elsewhere. The pharmaceutical industry is really two industries: the research-intensive companies that develop new drugs, and the manufacturing companies that make and sell existing unpatented drugs (also known as generics). By extending patents and preventing generic manufacture, the voucher would let the former group make more money for longer from the drugs they patented – and push the price of healthcare up. Whether that was justified or not depended on your perspective. Outterson, who once strongly opposed the idea of a transfer voucher on those grounds, now supports it for the same reason. He argues that since antibiotics serve as the foundation of modern medicine, it's morally justified and even desirable to spread the cost of new antibiotics over an entire healthcare system.

Legislation to persuade companies to develop unprofitable medicines has a chequered history. In 1982 in the US, similar reforms were made to improve the availability of 'orphan drugs' – the term for drugs for rare diseases which would be unprofitable because of the small number of people affected, defined in law as fewer than 200,000 patients. If a drug was awarded orphan status by the FDA, its developers won generous benefits similar to a voucher scheme, including seven extra years of market-exclusivity – even if the drug wasn't patentable – and substantial tax credits. Inevitably, companies manipulated this opportunity and created something paradoxical: orphan drugs that brought them billions of dollars in sales. A common tactic was to develop a drug that would be profitable for treating a common disease, then win designation as an orphan drug after finding a suitable rare disease. The drug could legally then be used for treating the common disease, as had always been envisaged, but with the added financial benefits conferred by its orphan status. The result was extravagant private profits that came from increasing the cost of routine healthcare: a 2021 study found that more than 70 per cent of spending on top-selling, multi-use orphan drugs was for treating common, rather than rare, diseases. It has been estimated that by 2028, the top ten orphan drugs will each bring in over $4 billion in sales, with a single company, Johnson & Johnson, expected to bring in $30 billion from the category.

The case of orphan drugs shows that attempts to stimulate private investment can end up costing far more than anticipated. The desire to avoid direct payments to companies means that governments are tempted to choose other options where the costs to taxpayers are ultimately higher, but better concealed. Antibiotics are valuable – but how valuable? The risk is that, in a panic, society pays far too much.

So far, our ability to design new antibiotics, like our ability to design new ways of developing them, has been underwhelming. The promise of AI is being publicised by pharmaceutical companies, such as in a partnership between Lilly and OpenAI to find new antibiotics announced in June 2024. But choosing the best molecules to invest in remains a struggle. It's possible that we are at the stage where drug developers could begin to apply some of the additional constraints we need new antibiotics to follow – not only killing bacteria, but also not killing us. Yet adding too many constraints may mean we miss precisely what we're searching for: the unexpected.

At the end of 2023, the MIT team behind halicin and abaucin

published another paper on a new antibiotic. This time, they did their best to add more constraints, using AI to search for molecules that killed bacteria but were predicted to be less toxic to humans. They filtered over 12 million molecules based on their predictions: only 10,000 had a high antibiotic prediction score; 3,600 of those had predicted low toxicity to human cells; and 2,200 of those were free of unfavourable chemical substructures which might play havoc with human physiology. Finally, the team found 1,200 molecules that seemed to have new structures when compared to existing antibiotics. They had used AI to filter down to less than 0.01 per cent of the chemical universe they'd started with.

The researchers then went one step further and tried to build an algorithm to explain *why* the molecules would work as antibiotics. Like the beta-lactam ring of penicillin, the algorithm would try to find the crucial substructure within: what the researchers called the 'rationale'. Inside the computer, the algorithm split the molecules apart many times into random fragments and predicted the antibiotic activity of each fragment, trying to find the crucial piece. The rationales it identified for known antibiotic classes matched what chemists already knew: for penicillins, it found the beta-lactam ring. The algorithm struggled to explain most of its predictions, only giving a predicted rationale for 15 per cent of molecules. But within those molecules, two that looked particularly promising shared a common rationale: a set of two benzene rings linked by a particular molecular bridge, reminiscent in its construction to the sulfa drugs. Perhaps that structure might eventually form the basis of a new antibiotic class.

Stokes, the Phare Bio co-founder who first identified halicin's antibiotic activity, wants to see antibiotic discovery heading in this direction. It's a hopeful glimpse of a use for AI beyond simply speeding up a process – expanding our minds instead. Stokes believes using AI should be iterative, a back-and-forth between computers and humans. He hopes AI can find more unexpected molecules in the chemical universe and teach us something genuinely new.

In the best case, it could be a little like the experience of human players of Go, the game that has been played for over 2,000 years on a simple board with black and white stones. Go's rules are childishly simple, but like chemical space or Borges's library, the universe of possible games is far too vast to explore in full. Every game is unique, with a layout of counters that has never been seen before and will never occur again. Go is far more complex than chess, at which computers surpassed humans in 1997, when IBM's Deep Blue defeated Garry Kasparov. Humans play Go based on heuristic rules and intuitions. How could a computer match that? In March 2016, the company DeepMind pitted a new computing system called AlphaGo against Lee Sedol, a Korean Go master widely considered the best player in the world. The month before, Lee had predicted that he would win in a landslide.

He was wrong. The world was shocked when AlphaGo won the first game. But there were more surprises in store. At the thirty-seventh move in the second game, AlphaGo played a completely unexpected move. A three-time European Go champion watching scribbled his initial thoughts: 'Here?! This goes beyond my understanding . . . A human would never dare play it!'

The move violated centuries of theory, ceding far too much territory on the board. Lee smiled at first, but as the minutes passed, his expression became increasingly serious. The score tilted in AlphaGo's favour. By the 211th move, Lee resigned. Afterwards, many experts focused on the thirty-seventh move as the game's crucial moment. The consensus was that although no human would ever have chosen the move, AlphaGo had somehow seen its value. It was a move of radical magic: incomprehensible yet beautiful.

One could say the same thing of evolution itself. Again and again, chemists have been baffled by nature's ingenuity. It's still happening: in 2025, a new study reported that human cells have a previously unknown ability to recycle old proteins into potent antibiotics, offering a potential source of new medicines that some researchers are optimistic could rival soil. But the amazement of scientists is nothing new. From the mystery of the beta-lactam ring to the unexpected bonds that aminoglycosides formed within themselves, evolution has advanced human understanding. Some have tried to obfuscate that contribution. Waksman, who won the Nobel Prize for streptomycin, maintained until his death the bizarre stance that antibiotics didn't exist in nature: he claimed that soil bacteria were unable to make antibiotics without 'our ingenuity in isolating and culturing them'. This was an argument made by a man desperate to justify himself, at a time when owning an antibiotic was a new and controversial development. Armed with streptomycin as a precedent, the pharmaceutical industry successfully altered patent law and enabled antibiotics to become intellectual property. But now, when the most valuable medicines to society have become the least valued, antibiotic development is caught in an economic bind. That is the problem at the heart of the crisis.

Conclusion

The art of healing

A conquest of this kind is never finished; the contingency remains.

Simone de Beauvoir

The Swiss city of Basel sits in one of Europe's most cosmopolitan corners. You can leave your house in Germany, walk to work in Switzerland, then have lunch in France. I'm visiting the city for a few days to attend a conference about antibiotic resistance.

I know Basel a little already. A few years ago, I spent several months here as a researcher at the Biozentrum, a seventy-three-metre tower of glass and steel that houses forty laboratories over fifteen stories. Despite being a visiting academic, I was given an office of my own. (A privilege I have never been given at any British university.) Even by Swiss standards, Basel-Stadt is rich – per capita by far the wealthiest of all of the country's cantons. The Rhine flowing through the middle of the city made it an ideal location for dye industries in the nineteenth century, which grew into the modern pharmaceutical industry. Two of the world's largest pharmaceutical companies – Roche and Novartis – still have their headquarters here. Yet Basel is not an ostentatious city; according to a German playwright, 'English understatement seems like megalomania when compared to that of the people of Basel.'

It wasn't always that way. Back in the early sixteenth century, Basel was a hotbed of radicalism. It was here in June 1527 that Paracelsus, a lecturer at the university, staged a public burning of some of the most revered works of medicine. Paracelsus defied conventions wherever he found them. He lectured in German rather than Latin, wearing an alchemist's leather apron rather than academic robes. His burning of books was a demonstration of his contempt for medical authorities like the medieval philosopher Avicenna or the ancient physician Galen. Paracelsus wrote, contemptuously, that his shoe-buckles contained more wisdom than both of them. 'If disease puts us to the test,' he wrote, 'all our splendour, title, ring, and name will be as much help as a horse's tail.' Rather than studying books, he thought that medicine should proceed from the world itself: 'The art of healing comes from nature, not from the physician. Therefore the physician must start from nature, with an open mind.' Nature gave hints in the form of plants and herbs – Paracelsus loved the folk remedies that most doctors sneered at – but he also thought that the cunning of the alchemist could make those treatments more powerful.

Paracelsus was no modern scientist. He believed in all manner of hokum, including gnomes that haunted mountains. But his conception of 'chemical medicine' was prescient and influential. After his death, his followers disdained learning about medicine solely from books. Instead, they wanted well-equipped laboratories to concoct new medicines from nature using chemical principles. The medicines they created were largely useless, but they didn't take this as evidence that Paracelsus was wrong. This ineffectiveness, they reasoned, was due to a lack of costly and exotic ingredients. Over time, the Paracelsian principle that cures could be found in nature became warped into a fixation on nature's rarest substances. The more impossible the ingredients were to obtain, the better the medicine must be. By the late seventeenth century, European nobles were being treated with such lavish concoctions as Goa stones, first made by Jesuit monks. A Goa stone was a melange of unimaginable decadence: coral, pearl, ruby, emerald and even narwhal tusk were ground into a paste and then moulded into a dense ball from which the patient would scrape delicate flakes to place on their tongue. Goa stones were purported to cure all ills, but were as useless as they were expensive. They were not unique in this regard. Until the twentieth century, if a doctor gave you a medicine you were usually better off if you only pretended to take it.

A Renaissance prince would have willingly parted with a sack of gold in exchange for the power inherent in a single dose of antibiotics. More than any other medicine, antibiotics are the spectacular realisation of Paracelsus's dream: substances present in nature which have been transmuted through chemistry.

The rise of antibiotic resistance is a nightmarish inversion of that dream, striking at the very heart of the Paracelsian alliance between nature and chemistry. The cures that nature provides, it can also take away. Perhaps the antibiotic era was not the dawn of a new age but only a brief aberration, an imbalance that is now being restored, as all things must. The several hundred people who have gathered at the Basel conference are agreed on one uneasy truth: antibiotics seem to be incompatible with the pharmaceutical industry as it currently exists – the industry that made this city rich.

If technocracy is a religion, Switzerland is its Mecca. Over the twentieth century it has managed to accumulate international organisations like trading cards, from the World Trade Office to the United Nations. This small landlocked country is the epicentre of global health, with the WHO headquarters in nearby Geneva. Inside the conference, speakers discuss the complexities of the global regulatory environment, legislative agendas and national action plans. It occurs to me that the scene at the conference is a little like the medieval city Paracelsus would have known. Employees of small startups dart around like street-sellers hawking their wares, hiding their desperation for capital behind forced smiles. Representatives from the giants of big pharma stroll around like powerful ambassadors from distant thrones, surrounded by gaggles of courtiers seeking favour. Policy experts and regulators gather like city burghers in hushed discussions, bowing their heads together to exchange solemn and urgent whispers.

As I sit in this air-conditioned room and make hurried notes, despite the forced enthusiasm of some of the speakers, the mood is subdued. I find myself thinking back to penicillin and the promise it once held of a world free from infectious disease. In 1960, two doctors would write that there was 'no reason why anyone on the face of the earth should have to suffer the ravages of such crippling diseases'. They believed that science had all the solutions.

Penicillin is an old antibiotic, but it is not outmoded. Although the majority of staphylococci worldwide are now resistant to penicillin, it remains a recommended treatment for many other infections. Those conditions include congenital syphilis, where pregnant women with syphilis transmit bacteria to the developing foetus in the womb; if the infection takes hold, the baby can be stillborn or die soon after birth. Injections with penicillin reduce transmission by 97 per cent; it is one of the most effective antibiotic therapies we know of.

Shockingly, congenital syphilis endures, and remains the second-leading cause of preventable stillbirth worldwide. The obstacle is not penicillin resistance but a lack of penicillin: one of the world's oldest antibiotics, which over sixty years ago already cost less than the bottle it came in, is not available to everyone. Millions of people in sub-Saharan Africa and Latin America lack access to penicillin, and every year, millions of people contract syphilis. According to the WHO, that lack of access means that over 50,000 babies die each year who could each have been saved by a penicillin treatment costing $2.

The failure to ensure access to penicillin is a terrible indictment of how wealthy countries discovered antibiotics and then discarded them. Penicillin shortages now even affect the Western world from time to time because of fragile global supply chains. According to an Al Jazeera report, during one such shortage in 2017, only four factories in the world were making penicillin, three of them in China. At the time of writing, the only remaining penicillin factory in the Western world is in the town of Kundl, Austria, converted from a brewery in the aftermath of WWII. In recent years, problems such as the increased price of sugar following the war in Ukraine have affected its ability to produce enough penicillin to meet demand, despite the fact that yields of penicillin are now over fifty times what they were at the end of WWII. Yet each year, tens of thousands of babies die for want of a medicine that costs less than a cup of coffee. Penicillin became so cheap that we forgot its real value.

What is a crisis? In English, 'crisis' was originally a medical word. It meant a crucial point in the course of a person's illness. The germ of that meaning was contagious, and 'crisis' began to spread and reproduce, the plight of a febrile patient a ghostly metaphor that hovered over all sorts of systems. Societies or situations, like people, could be sick and feverish – in crisis. Infectious disease outbreaks are where the term's original and derived meanings coalesce.

Few would disagree that the Covid-19 pandemic that enveloped the world in 2020 was a crisis in every sense of the word. By early 2021, the pharmaceutical industry had produced multiple effective vaccines using mRNA technology that had never been used in humans before. That rapid deployment amazed and shocked in equal measure, because it showed that something perceived as impossible had in fact been scientifically achievable for years; the problem was politics. That crisis was also an opportunity. Pfizer returned a net profit margin of over 20 per cent in 2021 thanks to sales of its vaccine, reminiscent of the 1950s glory days of the tetracycline cartel. For 2021 and 2022, Covid-19 remained the world's leading infectious cause of death, but by 2023 it had fallen out of first place. The Covid-19 crisis had abated. The diminution of Covid-19 restored another infectious killer to first place, a different respiratory disease: tuberculosis.

Each year, 10 million people fall ill with tuberculosis and over a million die. Tuberculosis is a devastating global pandemic, and drug-resistant forms mean that new antibiotics are urgently needed. Yet in 2020, tuberculosis research received less than 1 per cent of the funding invested in Covid-19. Understanding the crisis might seem too vast a challenge, one million deaths too large a number. So instead, let us take the course of one person's illness to stand for a society in crisis.

Almost sixty years after my grandfather entered the sanatorium at Grey Towers, Phumeza Tisile was nineteen and studying at university in South Africa in 2010 when she started to feel unwell. The first few doctors she visited didn't know what was wrong, but eventually an X-ray led to a diagnosis of pulmonary tuberculosis. She was forced to abandon her university studies to start receiving treatment: the standard cocktail of four antibiotics, including isoniazid, a cocktail that had been refined in the 1960s. But it didn't work. After five months, tests revealed she had a drug-resistant strain. She would need further antibiotics. Transferred to a new hospital, so weak she couldn't talk, Phumeza found herself lying alone in a dark room; she presumed the doctors thought she was already dead.

She was given older, more toxic antibiotics, one of which made her deaf, but the bacteria inside her were resistant to those, too. A year into treatment, she was finally diagnosed with extensively drug-resistant tuberculosis, known as XDR-TB. The only option was last-ditch therapy with high doses of yet more toxic antibiotics. They made Phumeza feel even sicker, but she wanted to live so she kept taking them. Doctors told her she had only a 20 per cent chance of surviving; one advised her to visit a priest and prepare for death. She was lucky: after two years of treatment, by August 2013, she was finally cured.

Phumeza would never forget the horrific nature of the antibiotics she had received. 'If I could change anything about having TB,' she said about her experience, 'it would be the treatment.' As a survivor she heard of a new antibiotic that could treat XDR-TB more quickly and effectively, without the terrible side-effects she had experienced. But there was a problem: it was protected by a patent.

Bedaquiline, an oral antibiotic approved in the US in 2012 while Phumeza was undergoing treatment, represented the first new class of antibiotic against tuberculosis in over four decades. In the world of tuberculosis, it was quickly hailed as a game-changer. But Johnson & Johnson (J&J) held a worldwide patent, controlling the entire global supply. Phumeza decided to campaign for access.

J&J had filed the original patent on bedaquiline in July 2003. The usual twenty-year expiry date for a patent meant that it was due to expire in 2023. But J&J was trying to extend its rights to bedaquiline in various countries by filing secondary patents – a technique known as 'evergreening'. It had deliberately filed a patent on a 'pharmaceutically acceptable' form of bedaquiline in 2007 for this purpose – despite mentioning this same form in the original patent application – and had lodged applications to extend the original patent until 2027 on these grounds.

To Phumeza, this was a moral outrage. Using the exclusivity of the original patent, J&J had struck different deals with different countries for bedaquiline supply, meaning that it could cost radically different amounts in different countries. In South Africa, where drug-resistant forms were a public health concern, J&J sold the drug to the government at more than twice its cost in some other countries. In India, of the 130,000 drug-resistant tuberculosis cases a year, only a few thousand were receiving bedaquiline. In 2019, the Indian government was still relying on small donations of bedaquiline from J&J through the US Agency for International Development rather than beginning its own procurement processes. Phumeza believed that neither governments nor J&J were acting fast enough to roll out bedaquiline, leaving patients with no option but the older generations of toxic antibiotics that had left her deaf – if they could even manage to obtain those.

In advance of the expiration of J&J's primary patent, Phumeza and another tuberculosis survivor, Nandita Venkatesan, filed a legal challenge in India to oppose J&J evergreening bedaquiline. They won. At the start of 2023, the Indian Patent Office declined J&J's application to prolong the patent. But elsewhere, J&J had fared better: back in South Africa, the patent had been successfully extended to 2027. As the July 2023 expiry date of the original patent approached, campaigners stepped up their efforts, including a boost from YouTuber John Green who encouraged his subscribers to bombard J&J with the demands. J&J then announced a deal to reduce the price of bedaquiline in many countries, but not all, and did not announce it would abandon its secondary patent attempts. Campaigners were cautiously optimistic about the deal but described it as 'a creative procurement solution' that failed to address the larger injustices. In September, the South Africa Competition Commission went a step further, launching an unprecedented investigation into J&J over its pricing of bedaquiline. J&J eventually agreed to reduce its price by 40 per cent and officially withdraw its secondary patents in South Africa, as well as promising not to enforce them in over 130 other countries.

These significant victories for public health were hard-won by Phumeza and other campaigners. The injustice of bedaquiline highlighted that the distribution of medicines was governed not by medical need but by a Kafkaesque tangle of national and international regulations. It had taken concerted effort to persuade governments to stop making polite appeals to J&J for access and get serious about using legislative powers. J&J's attempt to extend its patent was common practice in the industry. So was the fact that much of the research that bedaquiline emerged from had been publicly funded. Calculations by Treatment Action Group campaigners suggested that the public money invested in the clinical development of bedaquiline far exceeded J&J's own investment.

Antibiotic resistance is often described as a war with bacteria. But the inequalities of penicillin and bedaquiline expose an ugly truth. For many in the world's wealthiest countries, some of the world's worst infectious diseases are now quarrels in faraway places affecting people of whom they know nothing. The Covid-19 pandemic accelerated a revolution in vaccine technology, but data shows that, in the words of the Stop TB Partnership, 'twelve months of Covid-19 eliminated twelve years of progress in the global fight against tuberculosis'. The pandemic was a disaster for tuberculosis. It disrupted diagnosis, interrupted treatments and shifted attention away from the need for new treatments. And in early 2025, the US government announced an abrupt halt to millions of dollars of funding that disrupted tuberculosis response efforts around the world.

Phumeza's story is one of millions about antibiotic resistance that could be told today. She continues to campaign for fair access to antibiotics – for everyone. 'It never made sense to me that there are drugs out there but they are not easily available to those people who actually need them most,' she says.

*

My grandfather Philip survived both tuberculosis and Covid-19. Twice in his life he received treatments at the cutting edge of medicine, seventy years apart: in 1951, two new antibiotics; in early 2021, one of the first mRNA vaccines. In richer countries, the collective understanding of what was once an urgent war against tuberculosis and other bacterial diseases has decayed into a distant memory. My grandfather is one of its few surviving veterans. As antibiotic resistance rises around the world, that attitude is looking like a dangerous form of complacency. It is surely not a coincidence that as infectious diseases faded from view in the countries where most pharmaceutical research happens, research into new antibiotics slowed.

Since 1990, more than a million people each year have died from resistant infections. The OECD estimates that even within its high-income member states, one in five bacterial infections are resistant to antibiotics. If current trends continue, in a decade's time resistance rates to last-resort antibiotics will be twice as high as they were in 2005. Countries with ageing populations are in particular danger, because two-thirds of deaths are in people over sixty-five. In the British hospital where I once worked as a researcher, the proportion of bloodstream infections with *E. coli* that were resistant to amoxicillin and clavulanic acid (Augmentin) grew by more than 11 per cent a year between 1998 and 2016. Patients with resistant infections spend longer in hospital, placing greater pressure on already creaking healthcare systems. Within the EU, antibiotic resistance now means that each year patients spend a cumulative 32.5 million extra days in hospitals, equivalent to the acute bed capacity of Portugal's entire healthcare system. But these beds are scattered across a continent. The result of such a disparate problem is a plethora of commitments and plans, but little concrete action. At the conference in Basel, I'm told that over 120 countries have each written a National Action Plan for antibiotic resistance. Less than a quarter have actually allocated resources to it.

If we continue with the status quo, average global life expectancy will fall by around two years over the next decade. The crisis of antibiotic resistance demands new antibiotics. But even a radical new antibiotic like bedaquiline struggles for distribution. Access is not only a problem for poorer countries. At the conference, Brenda Waning from the Stop TB Partnership tells us that some of the old antibiotics now needed to treat multi-resistant tuberculosis are not available within the EU or the US – either because the antibiotics were never approved or because regulatory regimes have changed. She has taken phone calls from American parents desperate to get these antibiotics for their children. Resistant infections, wherever they emerge, can spread all around the world. We cannot tackle the antibiotic crisis unless we grapple with its political dimensions – and where they originated.

The system of patents that emerged over the twentieth century is killing antibiotics. Patents are not laws of nature, but a peculiar system of prizes that is meant to stimulate innovation. In the nineteenth century, it was believed to be unethical to patent medicines. Now, it's routine. But history reveals that another world is possible – because, once upon a time, it existed. Penicillin arose from years of basic research by academics; its production was scaled up by a massive government programme that encouraged open collaboration without a product patent. Isoniazid, also unprotected by a product patent, was developed to treat tuberculosis by multiple pharmaceutical companies because they saw the need to vanquish tuberculosis. The need for new antibiotics has not changed, but the pharmaceutical industry has. In 2012, Roche's Chief Patent Officer commented that an antibiotic like isoniazid 'would probably not be developed today'.

The pharmaceutical industry has produced vital medicines. But it is long past time to consider whether it is our only option for doing so. The public-private partnerships of CARB-X and GARDP are welcome developments, but they do not tackle the underlying assumption that the route to new medicines lies within the private sector. Currently, the world relies on the expertise in pharmaceutical companies to develop new antibiotics. That reliance has created the dangerous situation we find ourselves in today. When companies shelve their antibiotic programmes or go bankrupt, researchers lose their jobs, and the pool of available talent shrinks further. Antibiotic resistance is a global problem – perhaps it can only be satisfactorily addressed with a global solution.

More than any other type of medicine, antibiotics appear to be a powerful advocate for the status quo of pharmaceutical research. But that is an illusion – most were developed in a very different world. The pharmaceutical industry's assessment is that there is not a market for new antibiotics. Rather than constructing complex schemes to persuade companies back into the arena, the crisis could be an opportunity to adopt another approach.

The economist John Maynard Keynes once wrote that the important thing for government was 'not to do things which individuals are doing already, and to do them a little better or a little worse; but to do those things which at present are not done at all'. Developing new antibiotics at scale is one of those things. As a long-term vision, governments around the world could set up an international institute to develop antibiotics outside the private sector. As one group of academics suggested in 2020, such an effort could begin with the wholesale public purchase of all current commercial antibiotics in development. That would cost an estimated $5 billion: a large amount, but less than what would be needed to finance alternative mechanisms such as subscription models or prizes. Research could be directed to global healthcare priorities and gaps in existing treatments. Multiple analyses have concluded that collectively investing in new antibiotics, just like vaccine preparedness for future viral pandemics, would pay for itself many times over.

The development of most antibiotics rests on publicly funded research carried out in universities. At present, an academic scientist who discovers a promising new antibiotic must be prepared to dive into the murky waters of its commercialisation, about which most academics know nothing. Since 1980 in the US, the Bayh-Dole Act has meant that discoveries made with public funds can be claimed by the universities where they took place, rather than being automatically owned by the public. Any research that appears patentable ends up being conducted in an atmosphere of paranoia, with discoveries kept secret and publications delayed. The same approach is emulated in many other countries. In the case of antibiotics, this is particularly nonsensical, because the patents are unlikely to be commercially valuable: some researchers have even advocated for a completely 'open source' approach to antibiotic research that abandons patents altogether. Most academics do not do their research hoping to patent it. Where they do enter the confusing world of venture capital, it is to try and ensure their discoveries have a chance of reaching patients. An international antibiotics institute would provide an obvious alternative path. Universities worldwide could sign up to a charter to agree to waive their rights to intellectual property relating to antibiotics. The institute could provide legal support with filing patents in return for the associated rights, thus ensuring that the antibiotic entered into public ownership.

These antibiotics could still be manufactured and sold by companies under licence at fair prices, like under existing models such as the Global Drug Facility from the Stop TB Partnership or GARDP. But those companies would be generic manufacturers rather than research-intensive pharmaceutical companies. They would no longer need to recoup their investment in the antibiotic's development. Decoupling private sales from public research would remove the impossible demand for every antibiotic to be profitable under that model, and new antibiotics could be distributed based on need, rather than at the whim of a private patentholder. Access agreements between governments that funded the institute would recognise antibiotics as a common good, ensuring that they reach everyone who needs them. The inevitable emergence of resistance could be monitored with ongoing surveillance programmes, informing the most urgent directions for future development. The institute could be based at locations around the world, fostering an international community of antibiotic developers. By providing a long-term and stable home for researchers, it would strengthen the world's resilience in antibiotics, ensuring that each generation of researchers could pass on their knowledge to the next.

Such an international scientific collaboration has precedent. A similar approach has been used for other grand technological endeavours, from the Large Hadron Collider ($4.4 billion) to the International Space Station ($150 billion). Governments are prepared to collaborate to solve the deepest mysteries of particle physics and cosmology: why not antibiotics? There would be risks, including the certainty that some development programmes would fail to produce new medicines. The conventional model of pharmaceutical development outsources this risk to private companies – that is why antibiotic development has stalled. The calculus is entirely different for a public research programme, which doesn't need to worry about a return on investment in capital alone, because the improvements in health would be well worth the failures. Such a grand vision would not be expected to deliver results quickly: all the more reason to get started.

The biggest pharmaceutical corporations would no doubt protest that the government was infringing on their territory. Yet the past few decades have given them ample opportunity to stake their claim; the territory of antibiotics has been forfeited through inaction. As even drug developer John Rex admits, capitalism doesn't work for antibiotics. How could the public development of new antibiotics be in competition with an industry whose major players seem desperate to leave them behind? Quite the reverse: guaranteeing a stream of new antibiotics would make future medical innovations more possible. Antibiotics are so fundamental to modern medicine that they are almost a necessary condition of its existence. As the anthropologist Clare Chandler has put it, antibiotics are part of the infrastructure of modern society. Arguing that the state should steer clear of producing new antibiotics is a little like arguing it should stop building the roads that allow patients to get to hospitals. Indeed, since the pharmaceutical industry's profits rely on healthcare systems which in turn rely on antibiotics, a windfall tax on profits above a certain percentage could help fund the development of new antibiotics, keeping the industry's own markets sustainable.

The idea of a global effort to develop new antibiotics is not original. At the conference in Basel, when I put the suggestion to Kevin Outterson, the lawyer who founded CARB-X, he is polite but sceptical. Outterson points out that there is nothing comparable that is funded in the same way: when it comes to research and development, governments tend to pay private companies to do it. His own preferred approach is for governments to better reward the discovery and development of new antibiotics, while leaving it in the hands of the experts in the private sector who know how to do it.

Outterson and the other experts I talk to at the conference are sensible and dedicated professionals. They are well versed in operating within a landscape of constrained possibility, and are already working hard to achieve what they can within the status quo, buoyed up by recent alarm about antibiotic resistance. In a deeply dysfunctional system, many improvements are possible: CARB-X and GARDP are real achievements. Judging from my sense of the conference, there seems to be a growing coalition behind guaranteed revenue schemes where companies are paid a fixed amount per year for a new antibiotic, similar to the England subscription model, with each country paying for their fair share of the global problem. To me, such a proposal, while welcome, does not seem commensurate with the scale of the challenge. But to most at this conference, publicly owned antibiotic research sounds hopelessly utopian.

I wonder. Paul Ehrlich wanted to dye bacteria to death, his magic bullets a fantasy plucked from a folk tale. After hearing about Prontosil at a lecture in 1935, Alexander Fleming confided in a friend 'I've got something much better . . . but no one'll listen to me'. René Dubos claimed he could destroy disease-causing bacteria by experimenting with a cranberry bog. To turn these visions into reality took hundreds upon thousands of people: from Elva Akers, who allowed doctors to inject her with untested penicillin in wartime Oxford, to the scientists who screened soil, to the drug developers who continue to try to find new antibiotics today. But the success of antibiotics meant that what was once extraordinary has become commonplace. The unquestioned assumption is that antibiotics can be sustained by some version of business as usual.

Perhaps it is inevitable that we can only appreciate the value of antibiotics again once a world without them seems sufficiently close. The history of the last century suggests that exploitation comes naturally to humans. Those of us alive today have benefited from the past: from the efforts of the generations that came before us, and from the millions of years of evolution that made antibiotics possible at all. Reimagining antibiotics as a shared natural resource means reminding ourselves of our own responsibility to the future. The good news is that we have not yet reached the end of what we can learn from evolution. The challenge is to begin in earnest.

I go outside to my garden and pick up a handful of dark earth, crumbling it between my fingers. I watch it fall, knowing that among the millions of bacterial and fungal cells I cannot see, some are producing antibiotics that have saved millions of lives. It seems scarcely believable. Medicine from soil? Ridiculous.

Acknowledgements

First, my thanks to Seren Adams, who convinced me that I had something worth doing.

I am profoundly grateful to my agent Catherine Clarke, who understood and championed this book from the moment we met. Her wisdom and advice have made the process much easier. My editors Stuart Williams and Eamon Dolan could each be accurately described as peerless, were it not for the existence of the other: having their input and engagement has been a privilege. Thanks to Sam Wells for his excellent copy-editing. And thank you to everyone else who has worked on this book's production and distribution.

During the research for this book I had many valuable conversations with colleagues old and new, both in person and online. Thanks to them all. In particular, the experts in the history of antibiotics I reached out to were helpful and generous. Claas Kirchhelle gave reading suggestions and returned two early chapters with detailed comments. Karen Bush read several chapters and Robert Bud the whole manuscript in draft. All errors in what I have written remain my responsibility.

Thanks to Iain Pears who pointed me in the right direction then, much later, read the manuscript and knew what to say.

Thanks to those who I interviewed for specific details of the narrative: Ryan Cirz, Hank Heine, John Hollway, Kevin Judice, Heinz Moser, Kevin Outterson, Jon Stokes, Tim Walsh and Gerry Wright. To those I interviewed about subjects that didn't make it into the final book, apologies: Elisa Granato, Colin Kleanthous, Scott Matafwali and Jacob Palmer. For access to books and research documents I am grateful to Alexandra Elbakyan, the Internet Archive, and the staff at the Wellcome Collection and Bodleian Library.

Thanks to the many researchers whose work I have drawn on. I hope my appreciation for everyone who has contributed to both the science of antibiotics and the study of their history comes through in what I have written, as well as my admiration for those who have recognised the political and social dimensions of antibiotics. Solidarity with all who campaign for fair access to medicine.

While writing this book, my day job was researching bacterial evolution on a fellowship funded by the Wellcome Trust. Thanks to Craig MacLean (Oxford) for his support and mentorship. Thanks also to Richard Neher (Basel), Tim Blower (Durham), and Matthew Avison (Bristol), all of whom welcomed me into their research groups.

I am grateful to everyone who has given me opportunities to write. Thank you to the editors of the *Morning Star* and the *London Review of Books*. Particular thanks to Thomas Jones at the *LRB* blog, whose editing has taught me a great deal. Thanks also to *Science for the People*, who published an essay I wrote in 2022 about fossil drugs, sharpening the concept a lot in the process.

Thanks to my friends. In particular, since 2019, co-writing the 'Science and Society' column in the *Morning Star* every fortnight for six years with Rox Middleton, Joel Hellewell (until 2023), and Miriam Gauntlett (2023 onwards) has been one of the best collaborations I've been involved with, and has provided a valuable forum in which to think critically about science.

Thanks to Simon and Jo Hall, and to Beverley Ford, for their kindness and hospitality.

I owe a deep debt of thanks to my family. Thanks to my parents Cath and Quentin, who continue to inspire me with their actions to make the world a better place: I dedicate this book to them. Thanks to my brothers Pete, Luke and Joe for their friendship. And thanks to my grandfather, Phil, who graciously and characteristically gave his blessing to appear elsewhere in these pages.

My final and deepest expression of gratitude is to Claire Hall, who as well as being my favourite editor is (more importantly) the loveliest person I know. On my better days I *think* I could have written this book without your help. But not only would it have been much worse; I would have had much less fun. Wittgenstein once said 'I don't know why we are here, but I'm pretty sure that it is not in order to enjoy ourselves.' But he never met you.

Bibliography

For readers interested in more detail or alternatives, there are other books available. I mention here a few, working roughly backwards chronologically.

Miracle Cure (2017) by William Rosen is a highly readable account of the antibiotic era that finishes in the early 1960s. For an academic audience, Scott Podolsky's *The Antibiotic Era* (2015) is a rich history of attempts at antibiotic reform that is packed with archival colour. For antibiotics in agriculture, Claas Kirchhelle's *Pyrrhic Progress* (2020) is a comprehensive academic history of the US and Britain. For a general audience, Maryn McKenna's *Big Chicken* (2017) covers the rise of industrial farming and its dangers with characteristic acuity. McKenna has done a great deal to publicise antibiotic resistance, with her journalism reporting on real patient cases to bring its reality home: her coverage of MRSA in *Superbug* (2010) is the best book I know about a resistant bacterial threat. *Penicillin: Triumph and Tragedy* (2007) by Robert Bud encompasses much more than its title, with a second half covering attempts to tackle antibiotic resistance up to the mid-2000s. Though academic it is engaging and lively throughout. David Greenwood's *Antimicrobial Drugs* (2008) is a labour of love that provides a detailed chronology of antibiotics for a more scientific audience. For those who want to know more about the biology of resistance, *The Antibiotic Paradox* (1992) by Stuart Levy remains a reliable overview, with a second edition printed in 2002. Older works such as *Miracle Cure* (1990) by Milton Wainwright or Marc Lappé's *Germs that Won't Die* (1982) by now seem rather dated, though they contain some perceptive analysis.

Of the antibiotics that have a chapter in this book, I believe that only two have entire books devoted to them. For penicillin, Eric Lax's *The Mold in Dr. Florey's Coat* (2004) is a fast-paced and dramatic account that benefits from the close involvement of Norman Heatley. *Penicillin: meeting the challenge* (1985) by Gladys Hobby is a wonderful resource for any historian interested in its wartime production, but is difficult to get hold of. For streptomycin, Peter Pringle's *Experiment Eleven* (2012) is disturbing and compelling. Finally, anybody looking for a detailed history of the pharmaceutical industry should turn to Graham Dutfield's *That High Design of Purest Gold* (2020) as the new standard.

21st Century Cures Act (2016). *H.R. 34, 114th Congress.*

24/7 Wall St. (2018, May 3). *Why Achaogen Shares Are Getting Crushed.*

Abdurasulov, A. (2025, January 22). Dangerous drug-resistant bacteria are spreading in Ukraine. *BBC News.*

Abraham, E. P., & Chain, E. (1940). An Enzyme from Bacteria able to Destroy Penicillin. *Nature, 146*(3713), 837–837.

Acar, J. (1997). Broad- and narrow-spectrum antibiotics: an unhelpful categorization. *Clinical Microbiology and Infection, 3*(4), 395–396.

Achaogen (2013). NCT01970371: A Study of Plazomicin Compared With Colistin in Patients With Infection Due to Carbapenem-Resistant Enterobacteriaceae (CRE). *ClinicalTrials.Gov.*

Achilladelis, B. (1993). The dynamics of technological innovation: The sector of antibacterial medicines. *Research Policy, 22*(4), 279–308.

Adair, J., & Deuschle, K. W. (1970). *The people's health; medicine and anthropology in a Navajo community.* New York: Appleton-Century-Crofts.

Akiba, T. et al. (1960). On the mechanism of the development of multiple-drug-resistant clones of Shigella. *Japanese Journal of Microbiology, 4,* 219–227.

Alfano, A. (2016, January 11). *Solving the problem with penicillin.* Lab Dish (Cold Spring Harbor Laboratory). https://www.cshl.edu/labdish/solving-the-problem-with-penicillin/

Allasi Canales, N. (2022, April 7). *Hunting lost plants in botanical collections.* Wellcome Collection. https://wellcomecollection.org/articles/YjyPpREAAB8AhS-R

AMR Action Fund (2020, July 9). New AMR Action Fund steps in to save collapsing antibiotic pipeline with pharmaceutical industry investment of US$1 billion. *Business Wire.*

AMR Industry Alliance (2024). *Leaving the Lab: Tracking the Decline in AMR R&D Professionals.*

Anderson, D. (2016). *Animalcules.* Lens on Leeuwenhoek. https://lensonleeuwenhoek.net/content/animalcules

Andersson, D. I., & Hughes, D. (2011). Persistence of antibiotic resistance in bacterial populations. *FEMS Microbiology Reviews, 35*(5), 901–911.

Antunes, L. C. S., Visca, P., & Towner, K. J. (2014). Acinetobacter baumannii: evolution of a global pathogen. *Pathogens and Disease, 71*(3), 292–301.

Aronson, J. K., & Green, A. R. (2020). Me-too pharmaceutical products: History, definitions, examples, and relevance to drug shortages and essential medicines lists. *British Journal of Clinical Pharmacology, 86*(11), 2114–2122.

Asimov, I. (1972). *Asimov's Guide to Science.* New York: Basic Books.

Auerbacher, I. & Schatz, A. (2006). *Finding Dr. Schatz: the discovery of streptomycin and a life it saved.* New York: iUniverse.

Backus, M. P., Stauffer, J. F., & Johnson, M. J. (1946). Penicillin yields from new mold strains. *Journal of the American Chemical Society, 68*(1), 152–153.

Bailly, M. (2014, April 3). *Foolish Remedies: Goa stone.* Wellcome Collection. https://web.archive.org/web/20171201033146/https://wellcomecollection.org/articles/foolish-remedies-goa-stone

Baker, K. S. et al. (2015). The Murray collection of pre-antibiotic era *Enterobacteriacae*: a unique research resource. *Genome Medicine, 7*(1), 97.

Baldauf, S. L., & Palmer, J. D. (1993). Animals and fungi are each other's closest relatives: congruent evidence from multiple proteins. *Proceedings of the National Academy of Sciences of the United States of America, 90*(24), 11558–11562.

Baldry, P. (1976). *The Battle Against Bacteria: A Fresh Look.* Cambridge: Cambridge University Press. (Original work published 1965)

Ball, P. (2015, September 29). Navigating chemical space. *Chemistry World.* https://www.chemistryworld.com/features/navigating-chemical-space/8983.article

Barber, M. (1947a). Coagulase-positive staphylococci resistant to penicillin. *The Journal of Pathology and Bacteriology, 59*(3), 373–384.

Barber, M. (1947b). Staphylococcal infection due to penicillin-resistant strains. *BMJ, 2*(4534), 863–865.

Barber, M. (1960). "Celbenin"-resistant Staphylococci. *BMJ, 2*(5203), 939.

Barber, M. et al. (1960). Reversal of antibiotic resistance in hospital staphylococcal infection. *BMJ, 1*(5165), 11–17.

Barber, M. (1961). Methicillin-resistant staphylococci. *Journal of Clinical Pathology*, 14(4), 385–393.

Barber, M., & Dutton, A. A. C. (1958). Antibiotic-resistant staphylococcal outbreaks in a medical and in a surgical ward. *The Lancet*, 272(7037), 64–68.

Barber, M., & Rozwadowska-Dowzenko, M. (1948). Infection by penicillin-resistant staphylococci. *The Lancet*, 252(6530), 641–644.

BARDA (2023, April 27). *BARDA Support Protects Against Drug-Resistant Threats*. Medical Countermeasures. https://medicalcountermeasures.gov/stories/amr/

Bartlett, A. et al. (2022). A comprehensive list of bacterial pathogens infecting humans. *Microbiology*, 168(12), 001269.

Bastian, H. (2006). Down and almost out in Scotland: George Orwell, tuberculosis and getting streptomycin in 1948. *Journal of the Royal Society of Medicine*, 99(2), 95–98.

Beasley, D. (2019, April 23). Medicare offers to partially raise payment for cancer CAR-Ts. *Reuters*.

Becker, Z. (2024, April 24). *Orphan drug market to reach $270B by 2028, led by J&J, Vertex and Roche: Evaluate*. Fierce Pharma.

Beeson, P. B. (1990). Walsh McDermott: October 14, 1909-October 17, 1981. *Biographical Memoirs. National Academy of Sciences (U.S.)*, 59, 283–307.

Benedictow, O. J. (2004). *The Black Death, 1346-1353: The Complete History*. Woodbridge: Boydell Press.

Benzinga (2018, May 3). *Benzinga's Daily Biotech Pulse*.

Bergey, D. H. (1923). *Bergey's manual of determinative bacteriology*. Baltimore, MD: The Williams & Wilkins Company.

Best, W. R. (1967). Chloramphenicol-Associated Blood Dyscrasias: A Review of Cases Submitted to the American Medical Association Registry. *JAMA*, 201(3), 181–188.

Bickel, M. H. (1988). The development of sulfonamides (1932-1938) as a focal point in the history of chemotherapy. *Gesnerus*, 45 Pt 1, 67–86.

Binsker, U., Käsbohrer, A., & Hammerl, J. A. (2022). Global colistin use: a review of the emergence of resistant *Enterobacterales* and the impact on their genetic basis. *FEMS Microbiology Reviews*, 46(1), fuab049.

Blasco, B. et al. (2024). High-throughput screening of small-molecules libraries identified antibacterials against clinically relevant multidrug-resistant *A. baumannii* and *K. pneumoniae*. *eBioMedicine*, 102.

Blaser, M. J. (2014). *Missing Microbes: How the Overuse of Antibiotics Is Fueling Our Modern Plagues*. New York: Henry Holt & Company.

BMJ (1959, November 7). *Annotations*. 2(5157), 940–942.

BMJ (2019, May 9). *Indian campaigners ask government to stop relying on handouts of bedaquiline to ensure supply*. 365, l2140.

Bock, M. (2007). James Joyce and Germ Theory: The Skeleton at the Feast. *James Joyce Quarterly*, 45(1), 23–46.

Borges, J. L. (2001). *The total library: non-fiction 1922-1986*. London: Penguin Classics.

Bos, K. I. et al. (2011). A draft genome of *Yersinia pestis* from victims of the Black Death. *Nature*, 478(7370), 506–510.

Bosch, F., & Rosich, L. (2008). The Contributions of Paul Ehrlich to Pharmacology: A Tribute on the Occasion of the Centenary of His Nobel Prize. *Pharmacology*, 82(3), 171–179.

Boyer, B., Kroetsch, A., & Ridley, D. (2022). *Design of a Transferable Exclusivity Voucher Program*. Duke Margolis Center for Health Policy.

Bradford, P. A. et al. (2004). Emergence of carbapenem-resistant *Klebsiella* species possessing the class A carbapenem-hydrolyzing KPC-2 and inhibitor-resistant TEM-30 beta-lactamases in New York City. *Clinical Infectious Diseases* 39(1), 55–60.

Brasch, S. (2014, March 11). How China Became the World's Largest Pork Producer. *Modern Farmer*.

Breen, B. (2024, February 9). The (history of) spice must flow. *Res Obscura*. https://resobscura.substack.com/p/the-history-of-spice-must-flow

Brefeld, O. (1875). The Life-History of *Penicillium* (translation and abridgement of *Botanische Untersuchungen Tiber Schmimelpilze, Heft II*) (W. R. McNab, Trans.). *Journal of Cell Science*, s2-15(60), 342–359.

Brown, P. A. (2013). Diplomatic Farmers: Iowans and the 1955 Agricultural Delegation to the Soviet Union. *The Annals of Iowa*, 72.

Brunel, J. (1951). Antibiosis from Pasteur to Fleming. *Journal of the History of Medicine and Allied Sciences*, 6(3), 287–301.

Brust, J. C. M. (2023). 'Weighting' the Evidence: How Much Bedaquiline Is Enough? *American Journal of Respiratory and Critical Care Medicine*, 207(11), 1423–1424.

Buchanan, R. E. (1916). Studies in the Nomenclature and Classification of Bacteria: The Problem of Bacterial Nomenclature. *Journal of Bacteriology*, 1(6), 591–596.

Bud, R. (1993). *The Uses of Life: a history of biotechnology*. Cambridge: Cambridge University Press.

Bud, R. (2007a). *Penicillin: Triumph and Tragedy*. Oxford: Oxford University Press.

Bud, R. (2007b, January). Germ Warfare. *History Today, 57*(1), 30–32.

Buddhapia (2006, February 6). *About The Korean Tripitaka*. https://web.archive.org/web/20060206141217/http://www.buddhapia.com/buddhapi/news/campaign/haeinsa/e_p10.html

Burton, R. (2001). *The Anatomy of Melancholy*. New York: NYRB Classics. (Original work published 1621).

Bush, K., & Bradford, P. A. (2016). β-Lactams and β-Lactamase Inhibitors: An Overview. *Cold Spring Harbor Perspectives in Medicine, 6*(8), a025247.

Bush, K., & Bradford, P. A. (2020). Epidemiology of β-Lactamase-Producing Pathogens. *Clinical Microbiology Reviews, 33*(2), e00047-19.

Bush, L. M. et al. (2001). Index Case of Fatal Inhalational Anthrax Due to Bioterrorism in the United States. *New England Journal of Medicine, 345*(22), 1607–1610.

Bush, L. M., & Perez, M. T. (2012). The Anthrax Attacks 10 Years Later. *Annals of Internal Medicine, 156*(1), 41–44.

Business Monitor (2015, January 12). *Plazomicin QIDP Designation Highlights Achaogen As Acquisition Target*.

Business Monitor (2016, December 13). *Plazomicin Will Find Niche As Infections Adapt To Alternatives*.

Business Wire (2009a, February 3). *Achaogen Initiates Phase 1 Trial of ACHN-490 for Treatment of Multi-Drug Resistant Gram-Negative Bacterial Infections*.

Business Wire (2009b, March 3). *Achaogen Signs $26.6 Million Contract With NIAID for Development of New Therapy to Treat Resistant Strains of NIAID Category A and B Priority Pathogens*.

Business Wire (2010, August 30). *Achaogen Awarded Contract Worth up to $64 Million by BARDA for the Development of ACHN-490*.

Business Wire (2011, September 28). *Achaogen Appoints Chief Medical Officer Kenneth Hillan to Chief Executive Officer*.

CARB-X (2017). *CARB-X's first year: A story of global achievement and progress*. https://carb-x.org/carb-x-news/carb-xs-first-year-a-story-of-global-achievement-and-progress/

Carr, A., Stringer, J., & Shen, J. (2020). *Antibiotic and Antifungal Update: January 2020*. Needham.

Castanheira, M. et al. (2008). Rapid emergence of blaCTX-M among *Enterobacteriaceae* in U.S. Medical Centers: molecular evaluation from the MYSTIC Program (2007). *Microbial Drug Resistance, 14*(3), 211–216.

Catry, B. et al. (2015). Use of colistin-containing products within the European Union and European Economic Area (EU/EEA): development of resistance in animals and possible impact on human and animal health. *International Journal of Antimicrobial Agents, 46*(3), 297–306.

Cazares, A. et al. (2024). Pre and Post antibiotic epoch: insights into the historical spread of antimicrobial resistance. bioRxiv 2024.09.03.610986

CDC (2013). *Antibiotic resistance threats in the United States, 2013.*

Challenge Works (2024). *Transforming the approach to treating common infections: The PA-100 AST System - winner of the Longitude Prize on AMR.* Vimeo. https://vimeo.com/954315029

Chalmers, I. (2011). Why the 1948 MRC trial of streptomycin used treatment allocation based on random numbers. *Journal of the Royal Society of Medicine, 104*(9), 383–386.

Chan, J. F. et al. (2016). *Talaromyces (Penicillium) marneffei* infection in non-HIV-infected patients. *Emerging Microbes & Infections, 5*(1), 1–9.

Chan, P. F., Holmes, D. J., & Payne, D. J. (2004). Finding the gems using genomic discovery: antibacterial drug discovery strategies – the successes and the challenges. *Drug Discovery Today: Therapeutic Strategies, 1*(4), 519–527.

Chandler, C. I. R. (2019). Current accounts of antimicrobial resistance: stabilisation, individualisation and antibiotics as infrastructure. *Palgrave Communications, 5*(1), 53.

Chatterjee, S. (2021, March 11). The Long Shadow Of Colonial Science. *Noēma.*

Cheeseman, K. et al. (2014). Multiple recent horizontal transfers of a large genomic region in cheese making fungi. *Nature Communications, 5*(1), 2876.

Chen, J. C. et al. (2024). Stable isotope chemistry reveals plant-dominant diet among early foragers on the Andean Altiplano, 9.0–6.5 cal. ka. *PLOS ONE, 19*(1), e0296420.

Chen, K., & Wang, J. (2013, February 28). *Hog Farming in Transition: The Case of China.* The Pig Site. https://www.thepigsite.com/articles/hog-farming-in-transition-the-case-of-china

Chen, L. et al. (2021). Assessment of Mortality-Related Risk Factors and Effective Antimicrobial Regimens for Treatment of Bloodstream Infections Caused by Carbapenem-Resistant *Enterobacterales. Antimicrobial Agents and Chemotherapy,* 65(9), e00698-21.

Chester, F. D. (1901). *A manual of determinative bacteriology.* New York: The Macmillan company.

Chinese Ministry of Agriculture (2016, July 26). *Announcement No. 2428 of the Ministry of Agriculture regarding to the ban on the use of colistin sulfate for animal growth promoters.*

Chua, K.-P., Kimmel, L. E., & Conti, R. M. (2021). Spending For Orphan Indications Among Top-Selling Orphan Drugs Approved To Treat Common Diseases. *Health Affairs,* 40(3), 453–460.

Churchill, W. S. (1941, February 9). *'Give us the tools' address.*

Cirz, R. (2019, December 26). *J's Flu and the deal of the century.* https://www.linkedin.com/pulse/js-flu-deal-century-ryan-cirz/

Cirz, R. (2021, October 4). *When unmet medical need isn't enough: The Collapse of the Antibiotics Sector.* https://www.youtube.com/watch?v=RwxBhJScwXo

Clardy, J., Fischbach, M., & Currie, C. (2009). The natural history of antibiotics. *Current Biology,* 19(11), R437–R441.

Cluff, L. E. (1982). Walsh McDermott, 1909–1981. *The Journal of Infectious Diseases,* 146(2), 303–305.

Collier, R. (2014). The art and science of naming drugs. *Canadian Medical Association Journal,* 186(14), 1053.

Collins Lab (2020). *Harnessing AI to develop the first new class of antibiotics this generation.* The Audacious Project.

Congressional Budget Office (2021). *Research and Development in the Pharmaceutical Industry.*

Conway, E. M., & Oreskes, N. (2012). *Merchants of Doubt: How a Handful of Scientists Obscured the Truth on Issues from Tobacco Smoke to Global Warming.* New York: Bloomsbury.

Corley, T. A. B. (1994). The Beecham Group in the World's Pharmaceutical Industry 1914-70. *Zeitschrift Für Unternehmensgeschichte / Journal of Business History,* 39(1), 18–30.

Cornaglia, G. (2000). To the memory of an angel. *Clinical Microbiology and Infection,* 6, 1.

Corsello, S. M. et al. (2017). The Drug Repurposing Hub: a next-generation drug library and information resource. *Nature Medicine*, 23(4), 405–408.

Costello, P. M. (1968). The Tetracycline Conspiracy: Structure, Conduct and Performance in the Drug Industry. *Antitrust Law & Economics Review*, 1(4), 13–44.

Covell, G. (1955). Developments in the Chemotherapy of Malaria During the Last 40 Years. *Journal (Royal Society of Health)*, 75(10), 758–760.

Cramer, T. (2015). Building the 'World's Pharmacy': The Rise of the German Pharmaceutical Industry, 1871–1914. *Business History Review*, 89(1), 43–73.

Dale, H. (1949). Introduction. In M. Marquardt, *Paul Ehrlich*. London: W. Heinemann Medical Books.

Daniel, T. M. (2006). The history of tuberculosis. *Respiratory Medicine*, 100(11), 1862–1870.

Darwin, C. (1859). *On the origin of species by means of natural selection, or the preservation of favoured races in the struggle for life.* (1st ed.). London: John Murray.

Darwin, C. (1861). *On the origin of species by means of natural selection, or the preservation of favoured races in the struggle for life.* (3rd ed.). London: John Murray.

Datta, N., & Kontomichalou, P. (1965). Penicillinase synthesis controlled by infectious R factors in *Enterobacteriaceae*. *Nature*, 208(5007), 239–241.

Davies, M., & Walsh, T. R. (2018). A colistin crisis in India. *The Lancet Infectious Diseases*, 18(3), 256–257.

Dawkins, R. (2009). *The greatest show on Earth: the evidence for evolution.* London: Bantam.

D'Costa, V. M. et al. (2006). Sampling the antibiotic resistome. *Science*, 311(5759), 374–377.

D'Costa, V. M. et al. (2011). Antibiotic resistance is ancient. *Nature*, 477(7365), 457–461.

de Beauvoir, S. (1972). *The ethics of ambiguity* (B. Frechtman, Trans.). New York: Citadel Press. (Original work published 1947)

Deloitte (2019). *The rise of global health care companies: Applying best practices worldwide.*

Dickens, C. (1839). *Nicholas Nickleby.* London: Chapman & Hall Ltd.

DiMasi, J. A., Grabowski, H. G., & Hansen, R. W. (2016). Innovation in the pharmaceutical industry: New estimates of R&D costs. *Journal of Health Economics*, 47, 20–33.

Dimico, A., Isopi, A., & Olsson, O. (2017). Origins of the Sicilian Mafia: The Market for Lemons. *The Journal of Economic History*, 77(4), 1083–1115.

DNDi (2023, November 3). *Our story: 20 years of DNDi*. https://dndi.org/about/our-story-20-years-dndi/

Dobell, C. (1932). *Antony van Leeuwenhoek and his 'little animals'*. New York: Harcourt, Brace & Company.

Dodds, D. R. (2017). Antibiotic resistance: A current epilogue. *Biochemical Pharmacology*, 134, 139–146.

Donaldson, P. M. W., Palleti, A. P., & Carroll, M. P. (1994). Ciprofloxacin in general practice. *BMJ*, 308(6941), 1437.

Donev, J. (2024). *Energy density*. Energy Education. https://energyeducation.ca/encyclopedia/Energy_density

Dornberg, J. (1994, November 27). Basel, boring? Not on your life. This city is a Swiss mix of architecture, art and eateries. *Washington Post*.

Doyle, A. C. (1904). *A Study in Scarlet and the Sign of the Four*. London: Harper & Brothers.

Dubos, R. J., & Dubos, J. (1987). *The White Plague: Tuberculosis, Man, and Society*. Rutgers University Press. (Original work published 1952)

Dunleavy, K. (2023, April 18). Big Pharma Companies Ranked by Revenue 2022. *Fierce Pharma*.

Dutfield, G. (2020). *That High Design Of Purest Gold: A Critical History Of The Pharmaceutical Industry, 1880-2020*. World Scientific.

Dworkin, M., & Gutnick, D. (2012). Sergei Winogradsky: a founder of modern microbiology and the first microbial ecologist. *FEMS Microbiology Reviews*, 36(2), 364–379.

Ehrlich, P. (1908). *Experimental Researches on Specific Therapeutics*. London: H. K. Lewis.

Eisen, M. (2017). Patents are destroying the soul of academic science. *It Is NOT Junk*. https://www.michaeleisen.org/blog/?p=1981

Eisenstein, B. I., Oleson, F. B., & Baltz, R. H. (2010). Daptomycin: from the mountain to the clinic, with essential help from Francis Tally, MD. *Clinical Infectious Diseases:*, 50 Suppl 1, S10-15.

Elbert, W. et al. (2007). Contribution of fungi to primary biogenic aerosols in the atmosphere: wet and dry discharged spores, carbohydrates, and inorganic ions. *Atmospheric Chemistry and Physics*, 7(17), 4569–4588.

Elek, S. D., & Fleming, P. C. (1960). A new technique for the control of hospital cross-infection. Experience with BRL. 1241 in a maternity unit. *Lancet*, 2(7150), 569–572.

Eli Lilly. (2024, June 25). *Lilly collaborates with OpenAI to discover novel medicines to treat drug-resistant bacteria.* Lilly Investors.

Ellis, W. (2015, May 20). The Sea View Children's Hospital. *AbandonedNYC*. https://abandonednyc.com/2015/05/20/the-sea-view-childrens-hospital/

Endimiani, A. et al. (2009). ACHN-490, a Neoglycoside with Potent In Vitro Activity against Multidrug-Resistant *Klebsiella pneumoniae* Isolates. *Antimicrobial Agents and Chemotherapy*, 53(10), 4504–4507.

European Commission (2005). *Ban on antibiotics as growth promoters in animal feed enters into effect (IP/05/1687).*

European Medicines Agency (2016). *Updated advice on the use of colistin products in animals within the European Union (EMA/231573/2016).*

Eversana (2021, February 23). *AMR Action Fund Gets First CEO, Funding Boost.*

Fagin, D. (2013, March 22). Dye Me a River: How a Revolutionary Textile Coloring Compound Tainted a Waterway. *Scientific American.*

Falagas, M. E., & Kasiakou, S. K. (2006). Toxicity of polymyxins: a systematic review of the evidence from old and recent studies. *Critical Care*, 10(1), R27.

Farrar, J. (2019, April 21). We ignore the disaster in the antibiotics market at our peril. *Financial Times.*

Fay, I. W. (1919). *The Chemistry of the Coal-tar Dyes* (2nd ed.). New York: D. Van Nostrand Company.

FD Wire (2018, June 26). *Achaogen, Inc. Special Call - Final.*

FDA (2018, May 2). Antimicrobial Drugs Advisory Committee Meeting. *FDA Advisory Committee Hearings.* http://web.archive.org/web/20191214184001/https://www.fda.gov/media/114366/download

Federal Statistical Office, Switzerland (2023). *Cantonal gross domestic product (GDP) per capita 2008-2021* [Data set].

Ferreira Júnior, W. S. et al. (2012). Use and importance of quina (*Cinchona* spp.) and ipeca (*Carapichea ipecacuanha* (Brot.) L. Andersson): Plants for medicinal use from the 16th century to the present. *Journal of Herbal Medicine*, 2(4), 103–112.

Ferry, G. (2019). *Dorothy Crowfoot Hodgkin: Patterns, Proteins and Peace: A Life in Science.* London: Bloomsbury.

Fierce Biotech (2018, May 3). *Achaogen slides after 'yes and no' adcomm verdict on plazomicin.*

Finland, M., Jones, W. F., & Barnes, M. W. (1959). Occurrence of serious bacterial infections since introduction of antibacterial agents. *Journal of the American Medical Association, 170,* 2188–2197.

Finney, J. M. T., & Walker, G. (1917). Abrogate the patent on Salvarsan. *Journal of the American Medical Association, LXVIII*(21), 1572–1573.

Fischer, J., & Ganellin, C. R. (Eds.). (2006). *Analogue-based Drug Discovery.* Weinheim: Wiley-VCH.

Fleming, A. (1929). On the Antibacterial Action of Cultures of a *Penicillium*, with Special Reference to their Use in the Isolation of *B. influenzæ. British Journal of Experimental Pathology, 10*(3), 226–236.

Fleming, A. (1942). In-vitro tests of penicillin potency. *The Lancet, 239*(6199), 732–733.

Fleming, A. (1945, December 11). *Penicillin: Nobel Lecture.*

Fleming, A. (1946). *Chemotherapy: Yesterday, Today and Tomorrow.* Cambridge University Press.

Floersh, H. (2023, June 2). AI is helping find new antibiotics. Can they get to the clinic? *Fierce Biotech.*

Floudas, D. et al. (2012). The Paleozoic Origin of Enzymatic Lignin Decomposition Reconstructed from 31 Fungal Genomes. *Science, 336*(6089), 1715–1719.

The Fly (2016a, December 12). AKAO: *Recommendations.*

The Fly (2016b, December 13). AKAO: *Recommendations.*

Fontenay, C. L. (1980). *Estes Kefauver: a biography.* Knoxville, TN: University of Tennessee Press.

Foster, J. B., & Clark, B. (2020). *The Robbery of Nature: Capitalism and the Ecological Rift.* New York: NYU Press.

Frese, S. J. (2004, Winter). Comrade Kruschev and Farmer Garst. *Iowa Heritage Illustrated,* 146–153.

Frieden, T. R., & Mangi, R. J. (1990). Inappropriate use of oral ciprofloxacin. *JAMA, 264*(11), 1438–1440.

Fröhlich-Nowoisky, J. et al. (2009). High diversity of fungi in air particulate matter. *Proceedings of the National Academy of Sciences, 106*(31), 12814–12819.

Fromhage, L., & Houston, A. I. (2022). Biological adaptation in light of the Lewontin–Williams (a)symmetry. *Evolution, 76*(7), 1619–1624. https://doi.org/10.1111/evo.14502

Fry, S. (1991). *The Liar*. London: Heinemann.

Fuller, A. T. (1937). Is p-aminobenzenesulphonamide the active agent in Prontosil therapy? *The Lancet*, 229(5917), 194–198.

Gabriel, J. M. (2020). *Medical Monopoly: Intellectual Property Rights and the Origins of the Modern Pharmaceutical Industry*. University of Chicago Press.

Gagneux, S. (2018). Ecology and evolution of *Mycobacterium tuberculosis*. *Nature Reviews Microbiology*, 16(4), 202–213.

Gale, F., Marti, D., & Hu, D. (2012). *China's Volatile Pork Industry* (LDP-M-211-01). Economic Research Service, United States Department of Agriculture.

Gallagher, J. (2015, November 19). Antibiotic resistance: World on cusp of 'post-antibiotic era'. *BBC News*.

Gallagher, J. (2019, March 27). Take over pharma to create new medicines, says top adviser. *BBC News*.

Gardner, J. (2020, June 24). By $4M, La Jolla outbids Melinta for struggling Tetraphase. *BioPharma Dive*.

GARDP (2023, December 7). *GARDP webinar: An innovative public health approach to counter AMR*. https://www.youtube.com/watch?v=oOecomtOhFU

Garfield, S. (2001). *Mauve: how one man invented a color that changed the world*. London: W. W. Norton & Company.

Garrett, L. (2011, October 5). *Ten Years Ago Today the Anthrax Nightmare Unfolded, and Globalized*. Council on Foreign Relations. https://web.archive.org/web/20141006185045/http://www.cfr.org/public-health-threats-and-pandemics/ten-years-ago-today-anthrax-nightmare-unfolded-globalized/p26111

Garrod, L. P. (1966). Mary Barber. 3 April 1911—11 September 1965. *The Journal of Pathology and Bacteriology*, 92(2), 603–610.

Geist, K. L. (1978). *Pictures Will Talk: the life and films of Joseph L. Mankiewicz*. New York: Scribner.

Gelmo, P. (1908). Über Sulfamide der p-Amidobenzolsulfonsäure. *Journal Für Praktische Chemie*, 77(1), 369–382.

Genentech (2016, April 7). *Cloning Insulin*.

Gerber, N. N., & Lechevalier, H. A. (1965). Geosmin, an earthly-smelling substance isolated from actinomycetes. *Applied Microbiology*, 13(6), 935–938.

Gerlach, V. L. et al. (1999). Human and mouse homologs of *Escherichia coli* DinB (DNA polymerase IV), members of the UmuC/DinB superfamily.

Proceedings of the National Academy of Sciences of the United States of America, 96(21), 11922–11927.

Gigante, V. et al. (2024). Multi-year analysis of the global preclinical antibacterial pipeline: trends and gaps. *Antimicrobial Agents and Chemotherapy*, 0(0), e00535-24.

Gillespie, W. A., & Alder, V. G. (1957). Control of an outbreak of staphylococcal infection in a hospital. *The Lancet*, 269(6969), 632–634.

Gillings, M. R. (2014). Integrons: Past, Present, and Future. *Microbiology and Molecular Biology Reviews*, 78(2), 257–277.

Gilon, C. (2020, December 13). Indigo: The story of India's 'blue gold'. *Al Jazeera*.

Gingell, R., & Bridges, J. W. (1971). Intestinal azo reduction and glucuronide conjugation of prontosil. *The Biochemical Journal*, 125(2), 24P.

Gingell, R., Bridges, J. W., & Williams, R. T. (1971). The role of the gut flora in the metabolism of prontosil and neoprontosil in the rat. *Xenobiotica* 1(2), 143–156.

Gitzinger, M. (2022, October 26). *Designing SME-compatible pull incentives to reinvigorate innovation in AMR*. Workshop of ENVI Health Working Group.

Gitzinger, M. (2024, March 6). *2024, a decisive year: update on the AMR environment*. 8th AMR Conference, Basel.

Glasgow Medical Journal (1881). The International Medical Congress of 1881. *Glasgow Medical Journal*, 16(3), 185–191.

Global Leaders Group on AMR (2024). *GLG report: Towards specific commitments and action in the response to antimicrobial resistance.*

GlobeNewswire (2015, January 8). *Achaogen Announces Plazomicin Granted QIDP Designation by FDA.*

GlobeNewswire (2016, December 12). *Achaogen Announces Positive Results in Phase 3 cUTI and CRE Clinical Trials of Plazomicin.*

GlobeNewswire (2018a, January 2). *Achaogen Announces FDA Acceptance of New Drug Application with Priority Review for Plazomicin for Treatment of Complicated Urinary Tract Infections and Bloodstream Infections.*

GlobeNewswire (2018b, August 3). *CMS Grants New Technology Add-on Payment to ZEMDRI™ (plazomicin).*

Glover, R. E. et al. (2023). Why is the UK subscription model for antibiotics considered successful? *The Lancet Microbe*, 4(11), e852–e853.

Godfrey-Smith, P. (2017). The subject as cause and effect of evolution. *Interface Focus*, 7(5), 20170022.

Goldberg, K. et al. (2025). Cell-autonomous innate immunity by proteasome-derived defence peptides. *Nature*, 1–10.

Gonçalves, T., & Vasconcelos, U. (2021). Colour Me Blue: The History and the Biotechnological Potential of Pyocyanin. *Molecules*, 26(4), 927.

Gordon, A. (2023, July 18). Inside the Deal To Give Millions Access to Cheaper TB Drugs. *Time*.

Gorvett, Z. (2017, September 28). The deadly germ warfare island abandoned by the Soviets. *BBC Future*.

Gotham, D. et al. (2020). Public investments in the clinical development of bedaquiline. *PLOS ONE*, 15(9), e0239118.

Gradmann, C. (2011) Magic bullets and moving targets: antibiotic resistance and experimental chemotherapy, 1900-1940. *Dynamis* 31(2):305-21.

Grady, D. (2015, January 7). New Antibiotic Stirs Hope Against Resistant Bacteria. *New York Times*.

Graessle, O. E., & Pietrowski, J. J. (1949). The in vitro effect of para-aminosalicylic acid (PAS) in preventing acquired resistance to streptomycin by *Mycobacterium tuberculosis*. *Journal of Bacteriology*, 57(4), 459–464.

Gram, C. (1962). The differential staining of *Schizomycetes* in tissue sections and in dried preparations. In T. D. Brock, *Milestones in microbiology*. Englewood Cliffs, NJ: Prentice-Hall.

Graves, H. B. (1984). Behavior and Ecology of Wild and Feral Swine (*Sus Scrofa*). *Journal of Animal Science*, 58(2), 482–492.

Greenwood, D. (2008). *Antimicrobial Drugs: Chronicle of a Twentieth Century Medical Triumph*. Oxford: Oxford University Press.

Griffin, A. (2017, November 26). *See inside the abandoned Seaview Hospital & Farm Colony*. Staten Island Advance / Silive. https://www.silive.com/news/2017/11/a_look_inside_the_abandoned_se.html

Griffin, D. W. (2004). Terrestrial Microorganisms at an Altitude of 20,000 m in Earth's Atmosphere. *Aerobiologia*, 20(2), 135–140.

Griffiths, C. et al. (2004). Trends in MRSA in England and Wales: analysis of morbidity and mortality data for 1993-2002. *Health Statistics Quarterly*, 21, 15–22.

Grimbergen, A. J. et al. (2015). Microbial bet-hedging: the power of being different. *Current Opinion in Microbiology*, 25, 67–72.

Grimond, M. (1996, March 8). Ciba-Geigy and Sandoz to merge into £40bn giant. *The Independent*.

Groeger, L. V. (2012, April 4). A History of FDA Inaction on Animal Antibiotics. *ProPublica*.

Guimaraes, K. (2017, May 21). Why is the world suffering from a penicillin shortage? *Al Jazeera*.

Haensch, S. et al. (2010). Distinct Clones of *Yersinia pestis* Caused the Black Death. *PLOS Pathogens*, 6(10), e1001134.

Hager, T. (2006). *The Demon Under The Microscope*. New York: Harmony Books.

Hall, W., McDonnell, A., & O'Neill, J. (2018). *Superbugs: An Arms Race against Bacteria*. Cambridge, MA: Harvard University Press.

Halliday, T. (2022). *Otherlands: A World in the Making*. London: Penguin.

Hanif, M. O., Bali, A., & Ramphul, K. (2024). Acute Renal Tubular Necrosis. StatPearls.

Hansch, C. et al. (1962). Correlation of Biological Activity of Phenoxyacetic Acids with Hammett Substituent Constants and Partition Coefficients. *Nature*, 194(4824), 178–180.

Hare, R. (1970). *The birth of penicillin, and the disarming of microbes*. London: Allen & Unwin.

Hasan, S. (2022). *What's so special about Ukraine's black soil?* TRT World.

Hastings, M. (2010). *Overlord: D-day and the battle for Normandy 1944*. London: Pan Macmillan. (Original work published 1984)

Haucap, J., Rasch, A., & Stiebale, J. (2019). How mergers affect innovation: Theory and evidence. *International Journal of Industrial Organization*, 63, 283–325.

Hawthorne, F. (2003). *The Merck Druggernaut: The Inside Story of a Pharmaceutical Giant*. New York: John Wiley & Sons.

Hazen, R. M., & Morrison, S. M. (2022). On the paragenetic modes of minerals: A mineral evolution perspective. *American Mineralogist*, 107(7), 1262–1287.

Health News Institute (1958). *Facts about pharmacy and pharmaceuticals*. Health News Institute.

Hegyi, J. F. A. L. et al. (2007). *WO2008068231A1: Fumarate salt of (alpha s, beta r)-6-bromo-alpha-[2-(dimethylamino)ethyl]-2-methoxy-alpha-1-naphthalenyl-beta-phenyl-3-quinolineethanol*.

Hemarajata, P. (2019, June 17). *Revisiting the Great Imitator: The Origin and History of Syphilis*. American Society for Microbiology. https://asm.org:443/Articles/2019/June/Revisiting-the-Great-Imitator,-Part-I-The-Origin-a

Heron, G. A. (1882). Ehrlich's Method For The Detection Of Tubercle-Bacillus In Sputum. *BMJ*, 2(1137), 735–735.

Herzberg, D. (2013, October 15). Shadow Journals: The Story of Medical Advertising (Part 1 of 3). *Books, Health and History*. https://nyamcenterforhistory.org/2013/10/15/shadow-journals-part-1/

Hesse, W., & Gröschel, D. H. M. (1992). Walther and Angelina Hesse - Early Contributors to Bacteriology. *ASM News*, 58(8), 425–428.

Heyman, D. (2002). *Lessons from the Anthrax Attacks* (No. DTRA01-02-C–0013). Center for Strategic and International Studies.

Higgens, C. E., & Kastner, R. E. (1971). *Streptomyces clavuligerus* sp. nov., a β-Lactam Antibiotic Producer. *International Journal of Systematic and Evolutionary Microbiology*, 21(4), 326–331.

Higgins, S. H. (1919). *The Dyeing Industry* (3rd ed.). Manchester: Manchester University Press.

Hill, J. (2019). *How an Iowa Corn Farmer Befriended the Soviet Premier*. Tufts University.

Hillaby, J. (1952, July 2). Abuse of TB Drug Feared in Britain. *New York Times*, 27.

Hillan, K. (2014, September 9). *Testimony by Kenneth Hillan, CEO of Achaogen to House Energy and Commerce Subcommittee on Health Hearing*. Congressional Documents and Publications.

Hobby, G. L. (1985). *Penicillin: Meeting the Challenge*. Yale University Press.

Holland, T. L. et al. (2016). Infective endocarditis. *Nature Reviews. Disease Primers*, 2, 16059.

Holldorf, A. W. (2003). Roehl, Wilhelm. In *Neue Deutsche Biographie*. Bayerische Staatsbibliothek.

Hollway, J. (2010). Beyond venture capital. *Bioentrepreneur*, 1–3.

Honegger, R. et al. (2017). Fertile *Prototaxites taiti*: a basal ascomycete with inoperculate, polysporous asci lacking croziers. *Philosophical Transactions of the Royal Society B: Biological Sciences*, 373(1739), 20170146.

Honigsbaum, M. (2016). Antibiotic antagonist: the curious career of René Dubos. *Lancet*, 387(10014), 118–119.

Hotchkiss, R. D. (1990). From microbes to medicine: gramicidin, Rene Dubos, and the Rockefeller. In C. L. Moberg & Z. A. Cohn (Eds.), *Launching the Antibiotic Era*. New York: Rockefeller University Press.

Houbraken, J., Frisvad, J. C., & Samson, R. A. (2011). Fleming's penicillin producing strain is not *Penicillium chrysogenum* but *P. rubens*. *IMA Fungus*, 2(1), 87–95.

House of Lords Select Committee on Science and Technology (1998, March 17). *Science and Technology - Seventh Report*.

Howie, J. (1979). Gonorrhoea – a question of tactics. *BMJ*, 2(6205), 1631–1632.

Hughes, V. M., & Datta, N. (1983). Conjugative plasmids in bacteria of the 'pre-antibiotic' era. *Nature*, 302, 725–726.

Hui, F. (2016). *Match Review: Lee Sedol vs. AlphaGo. Move 37*. https://www.alphago-games.com/view/eventname/leesedol/game/1/move/37

Humphries, P. (2023). *Cleopatra and the Undoing of Hollywood: How One Film Almost Sunk the Studios*. Cheltenham: The History Press.

Hunermund, P., Schmidt-Dengler, P., & Takahashi, Y. (2014). Entry and Shake-out in Dynamic Oligopoly. *ZEW Discussion Paper No. 14-116*.

Hutchings, M. I., Truman, A. W., & Wilkinson, B. (2019). Antibiotics: past, present and future. *Current Opinion in Microbiology*, 51, 72–80.

Idsoe, O. et al. (1968). Nature and extent of penicillin side-reactions, with particular reference to fatalities from anaphylactic shock. *Bulletin of the World Health Organization*, 38(2), 159–188.

IFPMA (2024). *#AlwaysInnovating: The pharmaceutical innovation journey*. https://www.ifpma.org/initiatives/alwaysinnovating-the-pharmaceutical-innovation-journey/

Ikuta, K. S. et al. (2022). Global mortality associated with 33 bacterial pathogens in 2019: a systematic analysis for the Global Burden of Disease Study 2019. *The Lancet*, 400(10369), 2221–2248.

Irish Times (1952a, March 13). *A New Treatment for Tuberculosis*.

Irish Times (1952b, April 22). *News of the anti-T. B. drugs*.

Irving, D. (Ed.). (1983). *Adolf Hitler: the medical diaries: the private diaries of Dr Theo Morell*. London: Sidgwick & Jackson.

Jang, S., & Javadov, S. (2014). Inhibition of JNK aggravates the recovery of rat hearts after global ischemia: the role of mitochondrial JNK. PLOS ONE, 9(11), e113526.

Jessen, O. et al. (1969). Changing Staphylococci and Staphylococcal Infections. *The New England Journal of Medicine*.

Jimenez, D. (2022, February 9). Cutting the carbon footprint of pharma's supply chain. *Pharmaceutical Technology*.

Jin, K. et al. (2016). Total synthesis of teixobactin. *Nature Communications*, 7(1), 12394.

Jones, D. S. (2002). The health care experiments at Many Farms: the Navajo, tuberculosis, and the limits of modern medicine, 1952-1962. *Bulletin of the History of Medicine*, 76(4), 749–790.

Jones, E. (2001). *The business of medicine: the extraordinary history of Glaxo, a baby food producer, which became one of the world's most successful pharmaceutical companies*. London: Profile.

Jones, J. H. (1993). *Bad Blood*. New York: Free Press.

Journal of Cell Science (1881). Notes And Memoranda. s2-21(84), 650–654.

Journal Of The Society Of Chemical Industry (1916). The Cargo of the Submarine "Deutschland". *Journal Of The Society Of Chemical Industry*, 35(23).

Kadar, N. (2019). Rediscovering Ignaz Philipp Semmelweis (1818–1865). *American Journal of Obstetrics & Gynecology*, 220(1), 26–39.

Kaeberlein, T., Lewis, K., & Epstein, S. S. (2002). Isolating 'uncultivable' microorganisms in pure culture in a simulated natural environment. *Science*, 296(5570), 1127–1129.

Kaempffert, W. (1952, February 24). Science in Review: New Drugs That Combat Tuberculosis Hold Out A Promise of Far More Effective Control. *New York Times*.

Kamp, D. (1998, March 22). When Liz Met Dick: The Making of Cleopatra. *Vanity Fair*.

Kampfner, J. (2020, April 17). The forgotten French tapestry with lessons for our apocalyptic times. *Guardian*.

Kantha, S. S. (1991). A centennial review; the 1890 tetanus antitoxin paper of von Behring and Kitasato and the related developments. *The Keio Journal of Medicine*, 40(1), 35–39.

Kaplan, S. (2016, December 5). Winners and losers of the 21st Century Cures Act. *STAT*.

Kasumov, A. (2018, July 11). Novartis Exits Antibiotics Research, Cuts 140 Jobs in Bay Area. *Bloomberg*.

Kaufmann, S. H. E. (2008). Paul Ehrlich: founder of chemotherapy. *Nature Reviews Drug Discovery*, 7(5), 373–373.

Kenny, M., & Barber, M. (1944). Outbreak of Puerperal Sepsis due to a Single Type of *Streptococcus*. *BMJ*, 1(4354), 809–811.

Kevany, S. (2023, March 23). *E. coli* from meat behind half a million UTIs in the US every year, study suggests. *Guardian*.

Kevles, D. J. (2023, October 5). Unreasonable Terms. *New York Review of Books*.

Keynes, J. M. (1931). The End of Laissez-Faire (1926). In *Essays in Persuasion*. London: Macmillan.

Khrushchev, N. S., & Khrushchev, S. (2004). *Memoirs of Nikita Khrushchev: Commissar, 1918-1945*. Penn State Press.

Kieffer, N., Nordmann, P., & Poirel, L. (2017). *Moraxella* species as potential sources of MCR-like polymyxin resistance determinants. *Antimicrobial Agents and Chemotherapy*, 61(6), e00129-17.

Kingston, W. (2001). Innovation needs patents reform. *Research Policy*, 30(3), 403–423.

Kingston, W. (2004). Streptomycin, Schatz v. Waksman, and the Balance of Credit for Discovery. *Journal of the History of Medicine and Allied Sciences*, 59(3), 441–462.

Kinross, J. (2023). *Dark Matter: The New Science of the Microbiome*. Penguin Life.

Kirchhelle, C. (2018). Pharming animals: a global history of antibiotics in food production (1935–2017). *Palgrave Communications*, 4(1), 1–13.

Kirchhelle, C. (2020). *Pyrrhic Progress: The History of Antibiotics in Anglo-American Food Production*. New Brunswick, NJ: Rutgers University Press.

Kirchhelle, C. (2023). The Antibiocene – towards an eco-social analysis of humanity's antimicrobial footprint. *Humanities and Social Sciences Communications*, 10(1), 1–12.

Kirkpatrick, T. (2020, February 11). *Why the Most Dreaded Injection is Called the 'Peanut Butter' Shot*. Military.Com. https://www.military.com/off-duty/2020/02/10/why-most-dreaded-injection-called-peanut-butter-shot.html

Klein, E., Smith, D. L., & Laxminarayan, R. (2007). Hospitalizations and Deaths Caused by Methicillin-Resistant *Staphylococcus aureus*, United States, 1999–2005. *Emerging Infectious Diseases*, 13(12), 1840–1846.

Klug, D. M. et al. (2021). There is no market for new antibiotics: this allows an open approach to research and development. *Wellcome Open Research*, 6, 146.

Knapp, C. W. et al. (2010). Evidence of increasing antibiotic resistance gene abundances in archived soils since 1940. *Environmental Science & Technology*, 44(2), 580–587.

Knoll, A. H. (2011). The Multiple Origins of Complex Multicellularity. *Annual Review of Earth and Planetary Sciences, 39(39)*, 217–39.

Koenig, P. (1923). The First Submarine to Cross the Atlantic. In Horne, C. F. *Source Records Of The Great War: Vol. IV*.

Korea National Tourism Organization (2009). *Tripitaka Koreana at Haeinsa Temple*.

Kousoulis, A. A. (2012). Etymology of Cholera. *Emerging Infectious Diseases, 18(3)*, 540.

Laich, F., Fierro, F., & Martín, J. F. (2002). Production of Penicillin by Fungi Growing on Food Products: Identification of a Complete Penicillin Gene Cluster in *Penicillium griseofulvum* and a Truncated Cluster in *Penicillium verrucosum*. *Applied and Environmental Microbiology, 68(3)*, 1211–1219.

LaMotte, S. (2022, July 8). No antibiotics worked, so this woman turned to a natural enemy of bacteria to save her husband's life. *CNN*.

Landas, M. (2020). *Cold War Resistance: The International Struggle over Antibiotics*. University of Nebraska Press.

Landecker, H. (2016). Antibiotic Resistance and the Biology of History. *Body & Society, 22(4)*, 19–52.

Landmarks Preservation Commission (1985). *New York City Farm Colony - Seaview Hospital Historic District: Designation Report*.

Landsberg, G. M., & Denenberg, S. (2014). *Social Behavior of Swine*. MSD Veterinary Manual.

Lane, N. (2015). *The Vital Question: Why Is Life the Way It Is?*. London: Profile.

Lane, N. (2022). *Transformer: The Deep Chemistry of Life and Death*. London: Profile.

Lang, K. C. (2007). Chinese Tripiṭaka. In *Encylopedia of Buddhism*. London: Routledge.

Larsen, J. L., Bille, N., & Nielsen, N. C. (1973). Occurrence and Possible Role of *Moraxella* Species in Pigs. *Acta Pathologica Microbiologica Scandinavica Section B Microbiology and Immunology, 81B(2)*, 181–186.

Lasker, M. (1963, February 22). *Notable New Yorkers*. Columbia University Libraries Oral History Research Office.

Latour, B. (1988). *The Pasteurization of France*. Cambridge, MA: Harvard University Press.

Laurence, W. L. (1952, July 5). Effects on euphoria: Wide New Fields Seen for TB Drug, Including Aid to Narcotics Addicts. *New York Times*.

Laurence, W. L. (1961, March 12). Science: Drug for Staph. *New York Times*.

Lax, E. (2005). *The Mold in Dr. Florey's Coat: The Story of the Penicillin Miracle*. Henry Holt and Company.

Lear, J. (1959, January 3). Taking the Miracle out of the Miracle Drugs. *Saturday Review*, 42(1).

Leclerq, R. et al. (1988). Plasmid-Mediated Resistance to Vancomycin and Teicoplanin in *Enterococcus faecium*. *New England Journal of Medicine*, 319(3), 157–161.

Lederberg, J. (1988). Medical science, infectious disease, and the unity of humankind. *JAMA*, 260(5), 684–685.

Ledford, H. (2015, January 7). Promising antibiotic discovered in microbial 'dark matter'. *Nature*.

Lee, V. (2018). Microbial Transformations: The Japanese Domestication of Penicillin Production, 1946–1951. *Historical Studies in the Natural Sciences*, 48(4), 441–474.

The Legacy (2016, January 24). Wellcome Trust Has $6,461,000 Position in Achaogen Inc (AKAO).

Lehmann, J. (1949). The treatment of tuberculosis in Sweden with para-aminosalicylic acid; a review. *Diseases of the Chest*, 16(6), 684–703.

Lend-Lease Act (1941). H.R. 1776, 77th Congress.

Leroy, L. (1874, April 25). L'Exposition des impressionnistes. *Le Charivari*.

Lesch, J. E. (2007). *The First Miracle Drugs: How the Sulfa Drugs Transformed Medicine*. Oxford: Oxford University Press.

Levins, R., & Lewontin, R. (1985). *The dialectical biologist*. Cambridge, MA: Harvard University Press.

Levy, S. B. (2002). *The antibiotic paradox: how the misuse of antibiotics destroys their curative powers* (2nd ed.). Cambridge, MA: Perseus. (Original work published 1992)

Lewis, K. (2010, July). *The Uncultured Bacteria*. Small Things Considered. https://schaechter.asmblog.org/schaechter/2010/07/the-uncultured-bacteria.html

Lewis, S. (1925). *Arrowsmith*. New York: Harcourt, Brace & Company.

Li, J. et al. (2024). Deciphering the pivotal role of people with high-frequency occupational animal exposure in antibiotic resistance transmission between humans and animals. *The Journal of Antimicrobial Chemotherapy*, 79(1), 27–35.

Ling, L. L. et al. (2015). A new antibiotic kills pathogens without detectable resistance. *Nature*, 517(7535), 455–459.

Link, J. H. F. (1809). I. Observationes in Ordines plantarum naturales. *Der Gesellschaft Naturforschender Freunde Zu Berlin Magazin Für Die Neuesten Entdeckungen in Der Gesammten Naturkunde.*

Linton, D. S. (2010). War and Disease: Biomedical Research on Malaria in the Twentieth Century (review). *Journal of the History of Medicine and Allied Sciences, 65*(1), 143–145.

Lipinski, C. A. et al. (2001). Experimental and computational approaches to estimate solubility and permeability in drug discovery and development settings. *Advanced Drug Delivery Reviews, 46*(1–3), 3–26.

Lippi, D., & Gotuzzo, E. (2014). The greatest steps towards the discovery of *Vibrio cholerae. Clinical Microbiology and Infection, 20*(3), 191–195.

Liu, G. et al. (2023). Deep learning-guided discovery of an antibiotic targeting *Acinetobacter baumannii. Nature Chemical Biology, 19*(11), 1342–1350.

Liu, Y.-Y. et al. (2016). Emergence of plasmid-mediated colistin resistance mechanism *mcr-1* in animals and human beings in China: a microbiological and molecular biological study. *The Lancet Infectious Diseases, 16*(2), 161–168.

Llewelyn, M. J. et al. (2017). The antibiotic course has had its day. *BMJ, 358,* j3418.

Lloyd, D. G. et al. (2020). De Novo Resistance to Arg10-Teixobactin Occurs Slowly and Is Costly. *Antimicrobial Agents and Chemotherapy, 65*(1), 10.1128/aac.01152-20.

Lo, C. (2015, January 6). Pharma mergers: big business, bad science? *Pharmaceutical Technology.*

Locher, W. G. (2007). Max von Pettenkofer (1818–1901) as a pioneer of modern hygiene and preventive medicine. *Environmental Health and Preventive Medicine, 12*(6), 238–245.

Longitude Prize (2020, February 18). *Effectiveness of Cancer Treatments Threatened by Rising Antibiotic Resistance.* https://amr.longitudeprize.org/resources/effectiveness-of-cancer-treatments-threatened-by-rising-antibiotic-resistance/

Loots, D. T., van der Westhuizen, F. H., & Botes, L. (2007). Aloe ferox leaf gel phytochemical content, antioxidant capacity, and possible health benefits. *Journal of Agricultural and Food Chemistry, 55*(17), 6891–6896.

Lopatkin, A. J. et al. (2017). Persistence and reversal of plasmid-mediated antibiotic resistance. *Nature Communications, 8*(1), 1689.

Lowe, D. (2013, May 20). How Much Do Drug Companies Spend on R&D, Anyway? *Science: In the Pipeline.* https://www.science.org/content/blog-post/how-much-do-drug-companies-spend-r-d-anyway

Lowe, D. (2015, January 8). Teixobactin: A New Antibiotic From a New Platform? *Science: In the Pipeline.* https://www.science.org/content/blog-post/teixobactin-new-antibiotic-new-platform

Lozos, C. (2021, January 8). *Venrock's Bryan Roberts on the firm's new $450 million fund, and where it's shopping in 2021.* TechCrunch.

Ludden, C. et al. (2019). One Health Genomic Surveillance of *Escherichia coli* Demonstrates Distinct Lineages and Mobile Genetic Elements in Isolates from Humans versus Livestock. *mBio, 10*(1), e02693-18.

Macho, A. (2023). *Inside The Last Penicillin Factory In The West.* WorldCrunch.

MacInnes, C. (2011). *Absolute Beginners.* London: Allison & Busby. (Original work published 1959)

Maeder, T. (1994). *Adverse Reactions.* New York: Morrow.

Mahase, E. (2020). UK launches subscription style model for antibiotics to encourage new development. *BMJ, 369,* m2468.

Mahrer, A. (2023, May 26). How South San Francisco became a preeminent biotech hub. *San Francisco Business Times.*

Mai, J. et al. (2023). Fecal carriage and molecular epidemiology of *mcr-1*-harboring *Escherichia coli* from children in southern China. *Journal of Infection and Public Health, 16*(7), 1057–1063.

Maitland, H. (2022, November 24). The Mystery Of Elizabeth Taylor's 1961 Oscars Dress, Solved. *British Vogue.*

Malan, M. (2023, September 22). South Africa launches 'unprecedented' investigation of Johnson & Johnson over TB drug prices. *Guardian.*

Manchester Guardian (1952, April 4). *Caution in Use of Drug Urged.*

Mancini, I. et al. (2006). On the first polyarsenic organic compound from nature: arsenicin A from the New Caledonian marine sponge *Echinochalina bargibanti. Chemistry, 12*(35), 8989–8994.

Mann, T. (1927). *The Magic Mountain* (H.T. Lower-Porter, Trans.). New York: Knopf. (Original work published 1924)

Margulis, L., & Sagan, D. (1986). *Microcosmos: four billion years of microbial evolution.* New York: Summit Books.

MarketLine (2014, January 27). Achaogen raises $82.8 million in initial public offering. *Financial Deals Tracker.*

Marquardt, M. (1949). *Paul Ehrlich*. London: W. Heinemann Medical Books.

Martens, J. H. et al. (2002). Microbial production of vitamin B12. *Applied Microbiology and Biotechnology, 58*(3), 275–285.

Martin, Y. C. (2018). How medicinal chemists learned about log P. *Journal of Computer-Aided Molecular Design, 32*(8), 809–819.

Martinez, L. et al. (2022). Infant BCG vaccination and risk of pulmonary and extrapulmonary tuberculosis throughout the life course: a systematic review and individual participant data meta-analysis. *The Lancet Global Health, 10*(9), e1307–e1316.

Marx, K. (1890). *Capital (Volume 1)*. Progress Publishers.

Mayer, J. (2008, July 15). Excerpt: 'The Dark Side'. *NPR*.

McCarthy, M. W. (2019). Teixobactin: a novel anti-infective agent. *Expert Review of Anti-Infective Therapy, 17*(1), 1–3.

McCutchan, J. A., Adler, M. W., & Berrie, J. R. (1982). Penicillinase-producing *Neisseria gonorrhoeae* in Great Britain, 1977-81: alarming increase in incidence and recent development of endemic transmission. *BMJ (Clinical Research Ed.), 285*(6338), 337–340.

McDermott, W. (1969). The Story of INH. *The Journal of Infectious Diseases, 119*(6), 678–683.

McDermott, W., Deuschle, K. W., & Barnett, C. R. (1972). Health Care Experiment at Many Farms. *Science, 175*(4017), 23–31.

McDonnell, A. et al. (2024). *The Economics of Antibiotic Resistance*.

McFadyen, R. E. (1979). The FDA's regulation and control of antibiotics in the 1950s: the Henry Welch Scandal, Félix Martí-Ibáñez, and Charles Pfizer & Co. *Bulletin of the History of Medicine, 53*(2), 159–169.

McGlone, J. (2016). *The Crate: Its History and Efficacy*. Available at www.depts.ttu.edu/animalwelfare/research/sowhousing/documents/TheCrate.pdf

McGlone, J. J., & Salak-Johnson, J. (2009). *Changing from Sow Gestation Crates to Pens: Problem or Opportunity?* The Pig Site. https://www.thepigsite.com/articles/changing-from-sow-gestation-crates-to-pens-problem-or-opportunity

McGraw, D. J. (1976). *The antibiotic discovery era (1940-1960): vancomycin as an example of the era*. PhD thesis, Oregon State University.

McKenna, M. (2010). *Superbug: The fatal menace of MRSA*. New York: Simon & Schuster.

McKenna, M. (2017). *Big Chicken: The Incredible Story of How Antibiotics Created Modern Agriculture and Changed the Way the World Eats*. Washington, DC: National Geographic.

McKeown, T. (1979). *The Role of Medicine: Dream, Mirage, Or Nemesis?* Princeton: Princeton University Press.

McNeill, J. R. (2001). *Something New Under the Sun: An Environmental History Of The Twentieth Century World*. London: Penguin.

Medical Research Council (1950). Treatment of pulmonary tuberculosis with streptomycin and para-aminosalicylic acid; a Medical Research Council investigation. *BMJ*, 2(4688), 1073–1085.

Merker, M. et al. (2015). Evolutionary history and global spread of the *Mycobacterium tuberculosis* Beijing lineage. *Nature Genetics*, 47(3), 242–249.

Meselson, M. et al. (1994). The Sverdlovsk anthrax outbreak of 1979. *Science (New York, N.Y.)*, 266(5188), 1202–1208.

Miethke, M. et al. (2021). Towards the sustainable discovery and development of new antibiotics. *Nature Reviews Chemistry*, 5(10), 726–749.

Mietzsch, D. F., & Klarer, D. J. (1935). *Verfahren zur Herstellung von Azoverbindungen* (Patent No. DE607537C).

Mikami, K. (2019). Orphans in the Market: The History of Orphan Drug Policy. *Social History of Medicine*, 32(3), 609–630.

Moberg, C. L. (1999). René Dubos, a Harbinger of Microbial Resistance to Antibiotics. *Perspectives in Biology and Medicine*, 42(4), 559–580.

Moberg, C. L. (2005). *René Dubos, friend of the good earth*. Washington, DC: ASM Press.

Money Morning (2014, March 10). *IPO Calendar 2014: Four Companies Going Public This Week*.

Moody, E. R. R. et al. (2024). The nature of the last universal common ancestor and its impact on the early Earth system. *Nature Ecology & Evolution*, 1–13.

Morin, R. B. et al. (1962). Chemistry of Cephalosporin Antibiotics. I. 7-Aminocephalosporanic Acid from Cephalosporin C. *Journal of the American Chemical Society*, 84(17), 3400–3401.

Morris, W. (1893). Textiles. In *Arts and Crafts Essays*. London: Rivington, Percival & Company.

Morrison, B. (2010, September 3). Review of 'Under the Sun: The Letters of Bruce Chatwin' selected and edited by Elizabeth Chatwin and Nicholas Shakespeare. *Guardian*.

Morse, K. (2003). *The Nature of Gold: An Environmental History of the Klondike Gold Rush*. Seattle: University of Washington Press.

Moseley, P. et al. (2024). Resurgence of congenital syphilis: new strategies against an old foe. *The Lancet Infectious Diseases*, 24(1), e24–e35.

MSF (2014, November 17). *R&D cost estimates - MSF response to Tufts CSDD study on cost to develop a new drug*. MSF Access Campaign. https://msfaccess.org/rd-cost-estimates-msf-response-tufts-csdd-study-cost-develop-new-drug

MSF (2023, January 16). *India must reject Johnson & Johnson's attempt to extend monopoly on lifesaving TB drug bedaquiline*. MSF Access Campaign. https://msfaccess.org/india-must-reject-johnson-johnsons-attempt-extend-monopoly-lifesaving-tb-drug-bedaquiline

MSF (2024, July 5). *MSF and Health Justice Initiative welcome J&J's withdrawal of patents on lifesaving TB drug in South Africa*. MSF Access Campaign. https://www.msfaccess.org/msf-and-health-justice-initiative-welcome-jjs-withdrawal-patents-lifesaving-tb-drug-south-africa

Mukherjee, S. (2011). *The Emperor of All Maladies: A biography of cancer*. London: Fourth Estate.

Mukherjee, S. (2022, October 22). 'It is a flaw in our cells that becomes a flaw in love': doctor Siddhartha Mukherjee on the search for a cure for depression. *Guardian*.

Mulcahy, A. W. (2024). *Comparing New Prescription Drug Availability and Launch Timing in the United States and Other OECD Countries*. RAND Corporation.

Mulchandani, R. et al. (2023). Global trends in antimicrobial use in food-producing animals: 2020 to 2030. *PLOS Global Public Health*, 3(2), e0001305.

Munro, R. (1883). Minute Organisms and Their Relation to Disease: Part II. *Glasgow Medical Journal*, 19(6), 409–424.

Murray, C. J. L. et al. (2022). Global burden of bacterial antimicrobial resistance in 2019: a systematic analysis. *The Lancet*, 399(10325), 629–655.

Murray, G. G. R. et al. (2023). The emergence and diversification of a zoonotic pathogen from within the microbiota of intensively farmed pigs. *Proceedings of the National Academy of Sciences*, 120(47), e2307773120.

Murray, J. F. (2004). A Century of Tuberculosis. *American Journal of Respiratory and Critical Care Medicine*, 169(11), 1181–1186.

Naghavi, M. et al. (2024). Global burden of bacterial antimicrobial resistance 1990–2021: a systematic analysis with forecasts to 2050. *The Lancet*, 404(10459), 1199–1226.

Nagley, M. M. (1950). The combined use of paraaminosalicylic acid (PAS) and streptomycin in pulmonary tuberculosis. *Tubercle*, 31(7), 151–155.

National Office of Vital Statistics (1953). *Vital Statistics of the United States 1950 Volume III: Mortality Data*. U.S. Department of Health, Education, and Welfare.

Nelsen, M. P. et al. (2016). Delayed fungal evolution did not cause the Paleozoic peak in coal production. *Proceedings of the National Academy of Sciences*, 113(9), 2442–2447.

Neushul, P. (1993). Science, Government and the Mass Production of Penicillin. *Journal of the History of Medicine and Allied Sciences*, 48(4), 371–395.

New York Newsday (1990, January 30). *The Bacteria War (Part III: Discovery)*.

New York Times (1916, November 2). *The Deutschland eluded foe with $10,000,000 cargo*.

New York Times (1917, June 5). *Metz in Row at Capitol*.

New York Times (1923, June 26). *Metz admits paying for dye affidavits*.

New York Times (1939, September 17). *Sugar-Coated Germs*.

New York Times (1950, July 11). *John L. Smith dies; noted chemist, 61*.

New York Times (1953, January 29). *All-Out War on TB Held Possible Now*.

New York Times (1957, April 7). *TB drug is tried in mental cases*.

New York Times (1958, May 17). *Experts defend iproniazid's use*.

New York Times (1960, May 19). *U.S. Drug Aide got $287,142 on side*.

New York Times (1961a, February 18). *Bars 'Cleopatra' Plan: Fox Rejects Lloyd's Bid to Give Miss Monroe Lead*.

New York Times (1961b, March 5). *Elizabeth Taylor Undergoes Surgery*.

New York Times (1961c, March 6). *Miss Taylor Better but Still in Danger*.

New York Times (1961d, March 9). *Miss Taylor got Aid from Jersey: Woman Recommended Drug That Had Helped Her - Actress Improves*.

New York Times (1961e, March 28). *Elizabeth Taylor Here: Actress Will Continue Trip to Los Angeles Today*.

New York Times (1964, July 15). *Antibiotics plot charged to nine*.

New York Times (2001, October 12). *A Nation Challenged: Understanding Anthrax*.

NHS England (2022, June 15). *NHS lands breakthrough in global battle against superbugs* https://www.england.nhs.uk/2022/06/nhs-lands-breakthrough-in-global-battle-against-superbugs/

NICE (2012). *Assessing cost effectiveness*. https://www.nice.org.uk/process/pmg6/chapter/assessing-cost-effectiveness

Bibliography

NICE (2022). *Lessons learnt from the UK project to test new models for evaluating and purchasing antimicrobials: Report from external workshops (July and August 2022).*

Nichols, D. et al. (2010). Use of ichip for high-throughput *in situ* cultivation of 'uncultivable' microbial species. *Applied and Environmental Microbiology*, 76(8), 2445–2450.

Nikita Khrushchev Goes to Hollywood (1959). https://www.youtube.com/watch?v=e1uesFH20Bk

Nikita Khrushchev's Visit to Iowa (Film K-3056) (1959). Iowa State University Library University Archives.

Notegen, E. (2012). Introduction. In *Lifesavers for Millions* (pp. 8–13). Basel: Editiones Roche.

NovoBiotic (2020, June 16). *NovoBiotic Pharmaceuticals, LLC was awarded a 3-year, \$3MM SBIR Phase II NIH NIAID grant titled 'Teixobactin Development for Anthrax'.*

NovoBiotic (2023, April 25). *NovoBiotic Pharmaceuticals, LLC is awarded a new grant entitled 'Late-Stage Preclinical Development of Teixobactin to Treat Drug-Resistant Infections'.*

Noymer, A. (2007). Contesting the Cause and Severity of the Black Death: A Review Essay. *Population and Development Review*, 33(3), 616–627.

OECD. (2023). *Embracing a One Health Framework to Fight Antimicrobial Resistance.* OECD Health Policy Studies.

Ogunlana, L. et al. (2023). Regulatory fine-tuning of *mcr-1* increases bacterial fitness and stabilises antibiotic resistance in agricultural settings. *The ISME Journal*, 17(11), 2058–2069.

Okasha, S. (2024). Cancer and the Levels of Selection. *The British Journal for the Philosophy of Science*, 75(3), 537–560.

O'Neill, R. (2024, September 12). *Wars are breeding superbugs that will spread 'everywhere'.* Politico.

Onions, A. H. S. (1966). *Penicillium digitatum. Descriptions of Fungi and Bacteria,* Sheet 96.

Orbinski, J. (1999, December 10). *Doctors Without Borders Nobel Lecture.*

Ordooei Javan, A., Shokouhi, S., & Sahraei, Z. (2015). A review on colistin nephrotoxicity. *European Journal of Clinical Pharmacology*, 71(7), 801–810.

Osborne Industries (2021, June 3). The History of Pig Farrowing Pens. *Osborne Livestock Equipment.* https://osbornelivestockequipment.com/the-history-of-pig-farrowing-pens/

O'Shea, R., & Moser, H. E. (2008). Physicochemical Properties of Antibacterial Compounds: Implications for Drug Discovery. *Journal of Medicinal Chemistry, 51*(10), 2871–2878.

Outterson, K. (2009). The legal ecology of resistance: the role of antibiotic resistance in pharmaceutical innovation. *Cardozo Legal Review, 31,* 613.

Outterson, K. (2023, October 16). *WS 14 – Development of a Sustainable Market for New, Resistance-Breaking Antibiotics.* World Health Summit. https://www.youtube.com/watch?v=INHrIVahHUg

Pachter, H. M. (1951). *Paracelsus: Magic into Science.* New York: Henry Schuman.

Papagianni, M. (2007). Advances in citric acid fermentation by *Aspergillus niger*: Biochemical aspects, membrane transport and modeling. *Biotechnology Advances, 25*(3), 244–263.

Paracelsus (1988). Paracelsus 1493-1541, selected writings (J. Jacobi, Ed.; N. Guterman, Trans.). Princeton, NJ: Princeton University Press.

Parker, L. A. (1915). A Brief History of Materia Medica (Continued). *The American Journal of Nursing, 15*(9), 729–734.

The Paris Review (1990). *Iris Murdoch, The Art of Fiction No. 117. Summer 1990*(115).

Pathak, A. et al. (2020). Comparative genomics of Alexander Fleming's original *Penicillium* isolate (IMI 15378) reveals sequence divergence of penicillin synthesis genes. *Scientific Reports, 10*(1), 15705.

British Pathé (1961). Liz Wins Best Actress. YouTube. https://www.youtube.com/watch?v=9XL39_vrGMg

Päuser, S. (2012). Isoniazid (Rimifon): first specific against tuberculosis. In *Lifesavers for millions* (pp. 14–77). Basel: Editiones Roche.

Payne, D. J. et al. (2007). Drugs for bad bugs: confronting the challenges of antibacterial discovery. *Nature Reviews Drug Discovery, 6*(1), 29–40.

Pearce, R. M. (1912). Chance and the Prepared Mind. *Science, 35*(912), 941–956.

Percival, A. et al. (1976). Penicillinase-producing Gonococci in Liverpool. *Lancet,* 2(8000), 1379–1382.

Perrone, G., & Susca, A. (2017). Penicillium Species and Their Associated Mycotoxins. *Methods in Molecular Biology, 1542,* 107–119.

Pietsch, D. (2013). Krotovinas—soil archives of steppe landscape history. *Catena, 104,* 257–264.

Pitney, E. H., & Kasius, R. V. (1947). Tuberculosis Mortality in the United States and in Each State: 1945. *Public Health Reports (1896-1970), 62*(14), 487–511.

Pliny the Elder (1856). Book 35, An Account of Paintings and Colours. In John Bostock & H. T. Riley (Trans.), *The Natural History*.

Pobiner, B. (2013). Evidence for Meat-Eating by Early Humans. *Nature Education Knowledge*, 4(6).

Podolsky, S. H. (2010). Antibiotics and the Social History of the Controlled Clinical Trial, 1950–1970. *Journal of the History of Medicine and Allied Sciences*, 65(3), 327–367.

Podolsky, S. H. (2015). *The Antibiotic Era: Reform, Resistance, and the Pursuit of a Rational Therapeutics*. Baltimore, MA: Johns Hopkins University Press.

Podolsky, S. H. (2018). The evolving response to antibiotic resistance (1945–2018). *Palgrave Communications*, 4(1), 1–8.

Pogue, J. M. et al. (2011). Incidence of and risk factors for colistin-associated nephrotoxicity in a large academic health system. *Clinical Infectious Diseases* 53(9), 879–884.

Porter, R. (1998). *The Greatest Benefit to Mankind: a medical history of humanity*. New York: W. W. Norton & Company.

Powell, C. (2003, February 5). *Address to the U.N. Security Council*.

Pringle, P. (2012). *Experiment Eleven: dark secrets behind the discovery of a wonder drug*. London: Bloomsbury.

Project BioShield Act (2004). *108th Congress*.

Radden Keefe, P. (2021). *Empire of Pain: The Secret History of the Sackler Dynasty*. London: Picador.

Raju, T. N. (2000). The Nobel Chronicles. *The Lancet*, 355(9208), 1022.

Ramírez, C. (1986). A new species of *Penicillium* from the Chilean Tierra del Fuego. *Mycopathologia*, 96(1), 29–32.

Rapp, R. T. (2004). *Innovation Market Analysis After Genzyme-Novazyme*. NERA Economic Consulting.

Rappé, M. S., & Giovannoni, S. J. (2003). The uncultured microbial majority. *Annual Review of Microbiology*, 57, 369–394.

Reading, C., & Cole, M. (1977). Clavulanic acid: a beta-lactamase-inhibiting beta-lactam from *Streptomyces clavuligerus*. *Antimicrobial Agents and Chemotherapy*, 11(5), 852–857.

Reboux, G. et al. (2019). Identifying indoor air *Penicillium* species: a challenge for allergic patients. *Journal of Medical Microbiology*, 68(5), 812–821.

Rex, J. (2020, February 21). *Chemicals vs. drugs (Part 1): The end of bacitracin / the buzz around halicin.* AMR.Solutions. https://amr.solutions/chemicals-vs-drugs-the-end-of-bacitracin-the-buzz-around-halicin/

Rex, J. (2024). *Fire Extinguishers Of Medicine.* AMR.Solutions. https://amr.solutions/fire-extinguishers-of-medicine/

Rex, J., & Outterson, K. (2020, July 11). *Plazomicin EU marketing application is withdrawn: Near zero market value of newly approved antibacterials.* AMR. Solutions. https://amr.solutions/2020/07/11/plazomicin-eu-marketing-application-is-withdrawn-near-zero-market-value-of-newly-approved-antibacterials/

Reymond, J.-L. (2015). The Chemical Space Project. *Accounts of Chemical Research,* 48(3), 722–730.

Rieder, H. L. (1989). Tuberculosis among American Indians of the contiguous United States. *Public Health Reports,* 104(6), 653–657.

Riseman, N. J. (2007). 'Regardless of History'?: Re-Assessing the Navajo Codetalkers of World War II. *Australasian Journal of American Studies,* 26(2), 48–73.

Ritchie, H., Rosado, P., & Roser, M. (2024a). Fossil fuels. *Our World in Data.* https://ourworldindata.org/fossil-fuels

Ritchie, H., Rosado, P., & Roser, M. (2024b). Meat and Dairy Production. *Our World in Data.* https://ourworldindata.org/meat-production

Robinson, G. L. (1947). Penicillin in General Practice. *Postgraduate Medical Journal,* 23(256), 86–92.

Rocke, A. J. (1985). Hypothesis and experiment in the early development of Kekulé's Benzene theory. *Annals of Science,* 42(4), 355–381.

Rodengen, J. L. (1999). *The legend of Pfizer.* Fort Lauderdale, FL: Write Stuff Syndicate.

Røder, H. L. et al. (2021). Biofilms can act as plasmid reserves in the absence of plasmid specific selection. *NPJ Biofilms and Microbiomes,* 7(1), 1–6.

Romansky, M. J., & Rittman, G. E. (1944). A Method of Prolonging the Action of Penicillin. *Science* 100(2592), 196–198.

Rosen, W. (2018). *Miracle Cure: The Creation of Antibiotics and the Birth of Modern Medicine.* New York: Penguin.

Ross, R. (1902). *Researches on malaria: Nobel Lecture.*

Rzepa, H. (1996). *Mauveine: The First Industrial Organic Fine-Chemical.* Molecule of the Month. https://www.chm.bris.ac.uk/motm/mauveine/perkin.html

Sabih, A., & Leslie, S. W. (2024). Complicated Urinary Tract Infections. *StatPearls*.

Sampat, B. N. (2015). Intellectual property rights and pharmaceuticals: The case of antibiotics. *WIPO Economics & Statistics Series, Economic Research Working Paper No. 26*.

Sample, I. (2020, February 20). Powerful antibiotic discovered using machine learning for first time. *Guardian*.

San Millan, A. et al. (2017). Multicopy plasmids potentiate the evolution of antibiotic resistance in bacteria. *Nature Ecology & Evolution*, 1(1), 0010.

Sankaran, N. (Ed.). (2020). Diversifying the historiography of bacteriophages. In *Notes and Records: the Royal Society Journal of the History of Science* (Vols 74, issue 4).

Savoini, G. et al. (2002). Alternative antimicrobials in the nutrition of postweaning piglets. *Veterinary Record*, 151(19), 577–580.

Sayers, D. L. (1987). *The Mind of the Maker*. New York: Harper & Row. (Original work published 1941)

Schatz, A., Bugle, E., & Waksman, S. A. (1944). Streptomycin, a Substance Exhibiting Antibiotic Activity Against Gram-Positive and Gram-Negative Bacteria. *Experimental Biology and Medicine*, 55(1), 66–69.

Schlander, M. et al. (2021). How Much Does It Cost to Research and Develop a New Drug? A Systematic Review and Assessment. *Pharmacoeconomics*, 39(11), 1243–1269.

Schmitt, K., & Zacchia, N. A. (2012). Total decontamination cost of the anthrax letter attacks. *Biosecurity and Bioterrorism* 10(1), 98–107.

Schooley, R. T. et al. (2017). Development and Use of Personalized Bacteriophage-Based Therapeutic Cocktails To Treat a Patient with a Disseminated Resistant *Acinetobacter baumannii* Infection. *Antimicrobial Agents and Chemotherapy*, 61(10), 10.1128/aac.00954-17.

Schweinfurth, S. P. (2009). Chapter C: An Introduction to Coal Quality. In Brenda S. Pierce & Kristin O. Dennen (Eds.), *The National Coal Resource Assessment Overview: U.S. Geological Survey Professional Paper 1625–F*. https://pubs.usgs.gov/pp/1625f/

Scokeley (2014, June 2). *Here is Where: Penicillin Comes to Peoria*. HistoryNet. https://www.historynet.com/here-is-where-penicillin-comes-to-peoria/

Scott, P. T. et al. (2004). *Acinetobacter baumannii* Infections Among Patients at Military Medical Facilities Treating Injured U.S. Service Members, 2002--2004. *MMWR*, 53(45), 1063–1066.

Segar, S. (2023, November 22). *Interview: "Someone had to do it", says SA TB activist on Time 100 list*. Spotlight.

Selikoff, I. J., Robitzek, E. H., & Ornstein, G. G. (1952). Treatment of pulmonary tuberculosis with hydrazide derivatives of isonicotinic acid. *Journal of the American Medical Association, 150*(10), 973–980.

Shafiq, M. et al. (2023). Characterization of antibiotic resistance genes and mobile elements in extended-spectrum β-lactamase-producing *Escherichia coli* strains isolated from hospitalized patients in Guangdong, China. *Journal of Applied Microbiology, 134*(7), lxad125.

Shama, G., & Reinarz, J. (2002). Allied intelligence reports on wartime German penicillin research and production. *Historical Studies in the Physical and Biological Sciences, 32*(2), 347–367.

Shand, R. F., & Leyva, K. J. (2008). Archaeal Antimicrobials: An Undiscovered Country. In Paul Blum (Ed.), *Archaea: New Models for Prokaryotic Biology*. Poole: Caister Academic Press.

Shaw, L. (2023a, May 31). *Open the pod bay doors*. LRB Blog. https://www.lrb.co.uk/blog/2023/may/open-the-pod-bay-doors

Shaw, L. (2023b, July 20). *Evergreening*. LRB Blog. https://www.lrb.co.uk/blog/2023/july/evergreening

Sheehan, J. C. (1982). *The Enchanted Ring: the untold story of penicillin*. Cambridge, MA: MIT Press.

Sheldrake, M. (2020). *Entangled Life: How Fungi Make Our Worlds, Change Our Minds & Shape Our Futures*. London: The Bodley Head.

Shen, C. et al. (2020). Dynamics of *mcr-1* prevalence and *mcr-1*-positive *Escherichia coli* after the cessation of colistin use as a feed additive for animals in China: a prospective cross-sectional and whole genome sequencing-based molecular epidemiological study. *The Lancet Microbe, 1*(1), e34–e43.

Shen, C. et al. (2021). Prevalence of *mcr-1* in Colonized Inpatients, China, 2011–2019. *Emerging Infectious Diseases, 27*(9), 2502–2504.

Shen, Z. et al. (2016). Early emergence of *mcr-1* in *Escherichia coli* from food-producing animals. *The Lancet Infectious Diseases, 16*(3), 293.

Shwachman, H., & Schuster, A. (1956). The Tetracyclines: Applied Pharmacology. *Pediatric Clinics of North America, 3*(2), 295–303.

Silver, L. L. (2011). Challenges of Antibacterial Discovery. *Clinical Microbiology Reviews, 24*(1), 71–109.

Simpson, R. W. (1954). *Izoniazid in the Treatment of Schizophrenia.* PhD thesis, University of Glasgow.

Singer, A. C., Kirchhelle, C., & Roberts, A. P. (2020). (Inter)nationalising the antibiotic research and development pipeline. *The Lancet Infectious Diseases,* 20(2), e54–e62.

Sloan, D. (2006). *Geology of the San Francisco Bay Region.* Oakland, CA: University of California Press.

Smil, V. (2019). Energy (r)evolutions take time. *World Energy,* 44, 10–14.

Smith, J. L. (2010). Tushonka: Cultivating Soviet Postwar Taste. *M/C Journal,* 13(5).

Snesrud, E., McGann, P., & Chandler, M. (2018). The Birth and Demise of the IS*Apl1-mcr-1*-IS*Apl1* Composite Transposon: the Vehicle for Transferable Colistin Resistance. *mBio,* 9(1), e02381-17.

Snow, J. (1855). *On the mode of communication of cholera.* London: John Churchill.

Sontag, S. (1979). *Illness as metaphor.* New York: Vintage.

South San Francisco Historical Society (2020, June 22). *South San Francisco.* Conference of California Historical Societies.

Specht, T. et al. (2014). Complete Sequencing and Chromosome-Scale Genome Assembly of the Industrial Progenitor Strain P2niaD18 from the Penicillin Producer *Penicillium chrysogenum. Genome Announcements,* 2(4), 10.1128/genomea.00577-14.

Staley, J. T., & Konopka, A. (1985). Measurement of in situ activities of non-photosynthetic microorganisms in aquatic and terrestrial habitats. *Annual Review of Microbiology,* 39, 321–346.

Steen, K. (2014). Introduction. In K. Steen (Ed.), *The American Synthetic Organic Chemicals Industry: War and Politics, 1910-1930* University of North Carolina Press.

Steenwyk, J. L. et al. (2019). A Robust Phylogenomic Time Tree for Biotechnologically and Medically Important Fungi in the Genera *Aspergillus* and *Penicillium. mBio,* 10(4), 10.1128/mbio.00925-19.

Stewart, G. T. (1960). "Celbenin"-resistant Staphylococci. *BMJ,* 2(5205), 1085.

Stewart, G. T., & Holt, R. J. (1963). Evolution of Natural Resistance to the Newer Penicillins. *BMJ,* 1(5326), 308–311.

Stoker, N. (2023, June 30). *Bedaquiline and Tuberculosis (TB): What might the future hold? (full transcript).* Zenodo.

Stokes, J. M. et al. (2020). A Deep Learning Approach to Antibiotic Discovery. *Cell*, 180(4), 688-702.e13.

Stolberg, S. G. (2001, October 14). Anthrax Threat Points to Limits in Health System. *New York Times*.

Stop TB Partnership (2021, March 18). *12 Months of COVID-19 Eliminated 12 Years of Progress in the Global Fight Against Tuberculosis*.

Stop TB Partnership (2025, March 3). *Report on the Impact of US Government Funding Halt on TB Responses in High TB Burden Countries*.

Strathdee, S. A. et al. (2023). Phage therapy: From biological mechanisms to future directions. *Cell*, 186(1), 17–31.

Sullivan, B. (2022, March 11). Guns, not roses – here's the true story of penicillin's first patient. *The Conversation*.

Sun Tzu (1993). *The art of war* (Yuan Shibing, Trans.). Ware: Wordsworth Reference.

Swain, R. H. A. (1940). Strain Variations in the Resistance of *Streptococcus viridans* to Sulphonamide Compounds. *BMJ*, 1(4139), 722–725.

Sydney Morning Herald (1946, July 20). *Lourdes and Penicillin*.

Szreter, S. (2014). The Prevalence of Syphilis in England and Wales on the Eve of the Great War: Re-visiting the Estimates of the Royal Commission on Venereal Diseases 1913–1916. *Social History of Medicine*, 27(3), 508–529.

Tanner, Fred W. (1933). *Practical Bacteriology* (2nd ed.). New York: John Wiley & Sons. (Original work published 1928)

Taylor, N. P. (2020, January 2). Melinta files for bankruptcy in another dark day for antibiotics. *Fierce Biotech*.

Taylor, P. (2019, February 19). End of the line looms for Aradigm as it files for bankruptcy. *Fierce Biotech*.

Thiagarajan, K. (2023, March 28). India rejects application to extend patent on TB drug bedaquiline. *BMJ*, 380:p724.

Thompson, H. (1961, March 10). 'Cleopatra' held for Miss Taylor: Skouras Says Ill Star Will Appear in Film. *New York Times*.

Thompson, J. (2022, July 1). Life Helps Make Almost Half of All Minerals. *Quanta Magazine*.

Thornton, H. G. (1953). Sergei Nicholaevitch Winogradsky, 1856-1953. *Obituary Notices of Fellows of the Royal Society*, 8(22), 635–644.

Time (1929, January 28). *Science: The Garvans*.

Time (1944, January 10). *Admirable M&B*.

Time (1955, October 3). *The Good Wizard.*

Tisile, P. (2014, July 7). Meet the 23-year-old TB survivor taking on South Africa's patent laws. *Guardian.*

Tisile, P. (2021a, April 13). *The long and painful moments.* MSF. https://web.archive.org/web/20210413214341/https://blogs.msf.org/bloggers/phumeza/long-and-painful-moments

Tisile, P. (2021b, June 18). *The end of the journey?* MSF. https://web.archive.org/web/20210618155721/https://blogs.msf.org/bloggers/phumeza/end-journey

Tobbell, D. (2011). *Pills, Power, and Policy: The Struggle for Drug Reform in Cold War America and Its Consequences.* Oakland: University of California Press.

Tomlinson, C. (2021). *Tuberculosis Research Funding Trends, 2005–2020* (M. Frick, Ed.). Treatment Action Group.

Trading With The Enemy Act (1917). H.R. 4960, 65th Congress.

Trafton, A. (2020, February 20). *Artificial intelligence yields new antibiotic.* MIT News.

Traversa-Tejero, I. P. (2021). Estimation of the Population of Stars in the Universe. *Open Access Library Journal,* 8(12), 1–11.

Treatment Action Group (2023, July 13). *TAG Encouraged by J&J Deal to Reduce Bedaquiline Prices, but Company's Action Falls Short of Community Demands.* https://www.treatmentactiongroup.org/statement/tag-encouraged-by-jj-deal-to-reduce-bedaquiline-prices-but-companys-action-falls-short-of-community-demands/

Trevor-Roper, H. (1947). *The Last Days Of Hitler.* New York: Macmillan.

Tripathi, N., & Sapra, A. (2024). Gram Staining. *StatPearls.* StatPearls Publishing.

Umair, M. et al. (2023). International manufacturing and trade in colistin, its implications in colistin resistance and One Health global policies: a microbiological, economic, and anthropological study. *The Lancet Microbe,* 4(4), e264–e276.

UN (2024, September 9). *Political Declaration of the High-level Meeting on Antimicrobial Resistance.*

UNESCO (1995). *Haeinsa Temple Janggyeong Panjeon, the Depositories for the Tripitaka Koreana Woodblocks.* UNESCO World Heritage Centre.

University of Leicester (2023, May 16). *University launches pioneering new centre to study bacteriophages to combat antibiotic resistant bacteria*

US Congress (1919). September 22, 1919. In *Congressional Record: Proceedings and Debates of the United States Congress: Vol. 58 Part 6*. U.S. Government Printing Office.

US Fed News (2006, October 13). *Achaogen Wins $24.61 Million Contract*.

US Fed News (2007, June 21). *Achaogen Wins $18.79 Million Contract*.

Uttley, Anne H. C. et al. (1988). Vancomycin-resistant Enterococci. *The Lancet*, 331(8575–8576), 57–58.

Van Gestel, J. F. E. et al. (2003). *WO2004011436A1: Quinoline derivatives and their use as mycobacterial inhibitors*.

Van Leeuwenhoek, A. (1677). Observations, communicated to the publisher by Mr. Antony van Leewenhoeck, in a dutch letter of the 9th Octob. 1676. here English'd: concerning little animals by him observed in rain-well-sea- and snow water; as also in water wherein pepper had lain infused. *Philosophical Transactions of the Royal Society*, 12, 821–831.

Van Leeuwenhoek, A. (1683). Letter of 17 September 1683 to the Amsterdam: Royal Society, London. *Royal Society*, MS. 1898. L 1. 69.

Van Leeuwenhoek, A. (1939–2024). *Alle de Brieven / Collected letters*. Amsterdam: Royal Netherlands Academy of Sciences and Letters.

Vihta, K.-D. et al. (2018). Trends over time in *Escherichia coli* bloodstream infections, urinary tract infections, and antibiotic susceptibilities in Oxfordshire, UK, 1998–2016: a study of electronic health records. *The Lancet Infectious Diseases*, 18(10), 1138–1149.

Vysloužilová, B. et al. (2016). Chernozem. From concept to classification: a review. *AUC Geographica*, 51(1), 85–95.

Wainwright, M. (1990). *Miracle Cure: The Story of Penicillin and the Golden Age of Antibiotics*. Oxford: Basil Blackwell.

Wainwright, M. (1991). Streptomycin: discovery and resultant controversy. *History and Philosophy of the Life Sciences*, 13(1), 97–124.

Wainwright, M. (2004). Hitler's Penicillin. *Perspectives in Biology and Medicine*, 47(2), 189–198.

Wainwright, M. (2005). A Response to William Kingston, 'Streptomycin, Schatz versus Waksman, and the Balance of Credit for Discovery'. *Journal of the History of Medicine and Allied Sciences*, 60(2), 218–220.

Wakabayashi, D., & Fu, C. (2023, February 8). China's Bid to Improve Food Production? Giant Towers of Pigs. *New York Times*.

Waksman, S. A. (1945). *Microbial antagonisms and antibiotic substances*. New York: The Commonwealth Fund.

Waksman, S. A. (1946). Sergei Nikolaevitch Winogradsky: The Story of a Great Bacteriologist. *Soil Science*, 62(3), 197.

Waksman, S. A. (1954). *My Life With The Microbes*. New York: Simon & Schuster.

Waksman, S. A., & Lechevalier, H. A. (1949). Neomycin, a New Antibiotic Active against Streptomycin-Resistant Bacteria, including Tuberculosis Organisms. *Science*, 109(2830), 305–307.

Waksman, S. A., & Schatz, A. (1948). *Streptomycin and process of preparation* (Patent No. US2449866A).

Waksman, S. A., & Starkey, R. L. (1931). *The Soil and the Microbe*. New York: John Wiley & Sons.

Waksman, S. A., & Woodruff, H. B. (1940). The Soil as a Source of Microorganisms Antagonistic to Disease-Producing Bacteria. *Journal of Bacteriology*, 40(4), 581–600.

The Waksman Foundation of Japan (2017). *Report of Researches in 2014, 2015, 2016*.

Walker, J., & Tadena, N. (2012, December 31). J&J's Tuberculosis Treatment Sirturo Gains Approval. *Wall Street Journal*.

Wallerstein, R. O. et al. (1969). Statewide Study of Chloramphenicol Therapy and Fatal Aplastic Anemia. *JAMA*, 208(11), 2045–2050.

Wallman, I. S., & Hilton, H. B. (1962). Teeth pigmented by tetracycline. *The Lancet*, 279(7234), 827–829.

Walsh, T. R. et al. (2005). Metallo-beta-lactamases: the quiet before the storm? *Clinical Microbiology Reviews*, 18(2), 306–325.

Walsh, T. R. (2006). Combinatorial genetic evolution of multiresistance. *Current Opinion in Microbiology*, 9(5), 476–482.

Wang, R. et al. (2018). The global distribution and spread of the mobilized colistin resistance gene mcr-1. *Nature Communications*, 9(1), 1179.

Wang, X. (2020, October 8). Behind China's 'pork miracle': how technology is transforming rural hog farming. *Guardian*.

Wang, Y. et al. (2017). Comprehensive resistome analysis reveals the prevalence of NDM and MCR-1 in Chinese poultry production. *Nature Microbiology*, 2, 16260.

Wang, Y. et al. (2020). Changes in colistin resistance and *mcr-1* abundance in *Escherichia coli* of animal and human origins following the ban of colistin-positive additives in China: an epidemiological comparative study. *The Lancet Infectious Diseases*, 20(10), 1161–1171.

The War Office (1948). *Statistical Report on the Health of the Army, 1943-1945*. His Majesty's Stationery Office.

Ward, P. S. (1981). The American Reception of Salvarsan. *Journal of the History of Medicine and Allied Sciences, XXXVI*(1), 44–62.

Warner, M. (2013, August 15). *How horses helped cure diphtheria*. Smithsonian. https://americanhistory.si.edu/explore/stories/how-horses-helped-cure-diphtheria

Washington Post (1952, February 26). *KO For TB?*

Watanabe, T. (1963). Infective heredity of multiple drug resistance in bacteria. *Bacteriological Reviews*, 27(1), 87–115.

Watchlist News (2018, June 26). *Achaogen (AKAO) Sets New 12-Month Low at $8.73.*

Wechselmann, W., & Wolbarst, A. L. (Abraham L.) (1911). *The treatment of syphilis with salvarsan*. London: Rebman.

Weik, M. H. (1961). Clary Model DE 60. In *A Third Survey of Domestic Electronic Digital Computing Systems*. Ballistic Research Laboratories.

Weinstein, M. J. et al. (1970). Antibiotic 6640, a new *Micromonospora*-produced aminoglycoside antibiotic. *The Journal of Antibiotics*, 23(11), 551–554.

Welch, H., & Martí-Ibáñez, F. (1960). *The antibiotic saga*. New York: Medical Encyclopedia, Inc.

Wellcome Trust (2020). *The Global Response to AMR: Momentum, success, and critical gaps*.

Wells, N., Nguyen, V.-K., & Harbarth, S. (2024). Novel insights from financial analysis of the failure to commercialise plazomicin: Implications for the antibiotic investment ecosystem. *Humanities and Social Sciences Communications*, 11(1), 1–13.

The White House (2003, January 28). *President Delivers 'State of the Union'*.

WHO (2023). *WHO Model List of Essential Medicines*.

WHO (2017, February 27). *WHO publishes list of bacteria for which new antibiotics are urgently needed*. https://www.who.int/news/item/27-02-2017-who-publishes-list-of-bacteria-for-which-new-antibiotics-are-urgently-needed

Wielgoss, S. et al. (2011). Mutation Rate Inferred From Synonymous Substitutions in a Long-Term Evolution Experiment With *Escherichia coli*. *G3*, 1(3), 183–186

Williams, H. (2019, April 21). Could antibiotic-resistant 'superbugs' become a bigger killer than cancer? *CBS News*.

Winogradsky, S. (1924). La méthode directe dans l'étude microbiologique du sol. *Chimie et Industrie*, 11, 215–222.

Winters, M. (2024, July 11). Here's the median salary in the 25 biggest U.S. cities—see how you compare. *CNBC*.

Wistrand-Yuen, E. et al. (2018). Evolution of high-level resistance during low-level antibiotic exposure. *Nature Communications*, 9(1), 1599.

Womiloju, T. O. et al. (2003). Methods to determine the biological composition of particulate matter collected from outdoor air. *Atmospheric Environment*, 37(31), 4335–4344.

Wong, F. et al. (2024). Discovery of a structural class of antibiotics with explainable deep learning. *Nature*, 626(7997), 177–185.

Wong, S. S. (2016, September 8). Syphilis and the use of mercury. *The Pharmaceutical Journal*.

Woods, A. (2012). Rethinking the History of Modern Agriculture: British Pig Production, c.1910–65. *Modern British History*, 23(2), 165–191.

WHO (2022). *2021 antibacterial agents in clinical and preclinical development: an overview and analysis*.

Wright, S. S., & Finland, M. (1954). Cross-Resistance Among 3 Tetracyclines. *Proceedings of the Society for Experimental Biology and Medicine*, 85(1), 40–42.

Xconomy (2010, July 19). *Flush with cash, Achaogen continues antibiotic work*.

Yong, E. (2016). *I Contain Multitudes: The Microbes Within Us and a Grander View of Life*. London: The Bodley Head.

Young, J. P. W. (2016). Bacteria Are Smartphones and Mobile Genes Are Apps. *Trends in Microbiology*, 24(12), 931–932.

Zastrow, M. (2016, March 9). 'I'm in shock!' How an AI beat the world's best human at Go. *New Scientist*.

Zbar, A. P. (2022). *Syphilis: A Short Biography*. London: Springer International Publishing.

Zimmermann, M. et al. (2019). Mapping human microbiome drug metabolism by gut bacteria and their genes. *Nature*, 570(7762), 462–467.

Notes

Introduction
Fossil drugs

1   they dug holes: Morse 2003: 94–97

1   On a summer day in 2007: this and other details are drawn from D'Costa et al. 2011

2   a minority: see Bartlett et al. 2022

3   1 in 8 of all deaths: Ikuta et al. 2022

3   'no one recently qualified': quoted in Greenwood 2008: vii

3   one in five cancer patients: Longitude Prize 2020

4   These freezers contained: Baker et al. 2015

4   When scientists searched: Hughes & Datta 1983

4   In 2006: D'Costa et al. 2006

4   a common criticism: interview with Gerry Wright, 4 November 2022

5   an antibiotic called vancomycin: for details of vancomycin's discovery, see McGraw 1976. For the first modern reports of clinical resistance, see Leclerq et al. 1988; Uttley et al. 1988

6   history contains examples: see Sheldrake 2020: 9

6   Antibiocene: Kirchhelle 2023

6   the final second: The calculation is my own, based on the Earth's estimated age of 4.2 billion years. For a virtuoso compression of life's history into one year, see the opening pages of Yong 2016

6   'our competitors': Lederberg 1988.

7   a concerning picture: these statistics are from my own analysis based on the WHO Model List of Essential Medicines (WHO 2023) and the table from Fischer & Ganellin 2006: 490

7   In 2021, one survey: Miethke et al. 2021

7   In 2024, a report: Gigante et al. 2024

8   'one of the most urgent': UN 2024

8   a million deaths a year: C. J. L. Murray et al. 2022

8 By 2050, forecasts predict: Naghavi et al. 2024. The direct attribution to resistance comes by comparing to a counterfactual world in which those infections were susceptible to antibiotics.

8 Recent patient stories: all the people listed are members of the WHO Task Force of AMR Survivors – see https://www.who.int/groups/task-force-of-amr-survivors

9 Drug companies prefer: Collier 2014

9 a truly global history: Podolsky 2018 suggests that the first major effort to frame antibiotic resistance as a global problem was in 1981. There remains a great deal of important scholarship to be done on the history of antibiotics beyond the well-trodden paths of Anglophone academic history; I make no pretence that my book does anything of the sort.

Demons under the microscope
Germ theory

11 Demons under the microscope: the phrase is from Hager 2006

11 'The visible world': Margulis & Sagan 1986: 66

11 'my sole plague': Burton 2001: 12

11 may have killed over half: Benedictow 2004 estimates 60 per cent. The figure is disputed e.g. Noymer 2007

12 In vivid threads: Kampfner 2020

12 *Yersinia pestis*: Bos et al. 2011; Haensch et al. 2010

12 In the mid-1670s: Van Leeuwenhoek 1677. Leeuwenhoek describes first discovering 'living creatures in Rain water' in 1675 but only reports his observations from 1676.

12 vivid and madcap itinerary: for Leeuwenhoek's letters see Van Leeuwenhoek 1939-2024. Wine: Letter No. 82 [43] 5 January 1685, frogspawn: Letter No. 110 [65] 7 September 1688, tulips: Letter No. 88 [47]. October 12th 1685, snails: Letter No. 188 [110] 10 September 1697.

12 'very extravagantly': Van Leeuwenhoek 1683: 568

12 observations were so accurate: see Dobell 1932

12 'little animals': as Douglas Anderson makes clear, this is closer to the sense of his Dutch term *dierkens* than 'animalcules' – see Anderson 2016

12 'Men in a kingdom': Van Leeuwenhoek 1683: 570

13 'formless worms': Lane 2015: 9

13 'every naturalist . . . beautiful organisation': Darwin 1861: 135. Darwin added this material in the third edition.

14 enzymes are at work: for example, see Figure 2 of Gerlach et al. 1999. In the letters scientists use to describe protein sequence, one shared stretch runs AAVEMRDNP.

15 In the thirteenth century: Lang 2007: 759

15 white birch: Buddhapia 2006

15 soaked in seawater: details from Korea National Tourism Organization 2009

15 security guards and a fire truck: UNESCO 1995

15 long-term experiment: Wielgoss et al. 2011. The substitution rate they observed can be used to calculate the number of mutations in 1,000 generations ($8.9 \times 10^{-11} \times 4.63 \times 10^6 \times 1,000 = 0.4$).

16 The first life: my discussion here is necessarily speculative, but I find the suggestion that the first cells arose before genetic information more convincing than the hypothesis of an RNA world. See Lane 2022

16 'The entire bacterial world': quoted in Landecker 2016

17 'a black hole': Lane 2015: 1

17 flourished into multicellular life: bacteria have forms of multicellularity such as biofilms, where cells live in a structured community. But complex multicellularity is much rarer. It has emerged several times but, as far as we know, only in eukaryotes. See Knoll 2011

18 'a pathological mirror': Siddhartha Mukherjee 2011: 192

18 cellular atavism: for a discussion of cancer and the evolution of multicellularity, see Okasha 2024

19 infective endocarditis: details of pathogenesis taken from Holland et al. 2016

19 Greek word: Kousoulis 2012

19 'It travels': Snow 1855: 2

19 the worst that he had ever seen: 'the most terrible', Snow 1855: 38

20 'I found so little impurity': Snow 1855: 39

20 'all the more disturbing': Latour 1988: 33

20 'One obtains': quoted in Latour 1988: 81

20 never seen in nature: see Latour 1988: 82

20 a needle-thin glass rod: as quoted in Munro 1883: 415

21 *Vibrio cholerae:* the bacterium had been previously identified by Filippo Pacini in 1854, but Koch was unaware of this discovery. See Lippi & Gotuzzo 2014

21 Pettenkofer: Locher 2007

21 Koch demonstrated: Journal of Cell Science 1881

22 four hours: Glasgow Medical Journal 1881: 189

22 'visible to the naked eye': Glasgow Medical Journal 1881: 191

22 Fanny Hesse: Hesse & Gröschel 1992

22 an 'impression' of a scene: Leroy 1874

23 'far from being': Latour 1988: 26

23 pyocyanase: see Gonçalves & Vasconcelos 2021

23 In 1890: Kantha 1991

24 bled nine times: Warner 2013

The red thread
Prontosil

25 'Without the thought': Sayers 1987: 29

25 swampy forests: see Halliday 2022: 196

26 flammable black rock: some scientists believe coal formed because the ability of microorganisms to break apart lignins only evolved millions of years later e.g. Floudas et al. 2012. Nelsen et al. 2016 disagree.

26 The first stage: this discussion is based on the US Geological Survey's coal ranking system. See Schweinfurth 2009

26 ward off snakes: Pliny the Elder 1856: chap. 34

27 50 per cent greater: Table 1 of Donev 2024

27 main source of energy: calculated comparing 1850 to 2020 using data from Ritchie, Rosado & Roser 2024a. The transition was slow. It was only by 1884 that the majority of primary energy in the US came from coal rather than biomass. See Smil 2019

27 the first antibiotics: my application of the term here is anachronistic.

27 'a gigantic ally': Ross 1902

28 Cinchona: for an overview see Ferreira Júnior et al. 2012, although many aspects of the history of quina are disputed.

28 a far greater demand for quinine: for the story of exploitation behind *Cinchona ledgeriana*, planted in huge numbers by the Dutch in Indonesia, see Chatterjee 2021 and Allasi Canales 2022.

28 'happy experiment': quoted in Garfield 2001: 33. Quinine's elemental composition was known to be $C_{20}H_{22}N_2O_2$. Chemists hoped it could be produced from other compounds. For example, starting with naphthalene ($C_{10}H_8$), one could add an amine group (-NH_2) to form naphthalidine ($C_{10}H_9N$). Relative to quinine, this was lacking oxygen and hydrogen. Water (H_2O) was an obvious source of both. If – and it was a big if – one could combine a mixture of two naphthalidines with two waters, the elemental arithmetic would produce something tantalisingly close to quinine.

29 His experiments to produce quinine: see Garfield 2001: 35–37

29 mauveine: Rzepa 1996

29 sea snail: Garfield 2001: 40

29 'slime that adheres': Pliny the Elder 1856: chap. 27

29 economically important: Gilon 2020

29 'most wonderful and most useless': Morris 1893

29 black and repulsive: 'black, sticky, fetid, repulsive', quoted in Rocke 1985

29 Concentrated chemistry: not all the dyes listed here were discovered in the nineteenth century e.g. Alcian blue was discovered in 1947.

30 By 1900: Fay 1919: 4

30 cigars and mineral water: 'strong cigars and mineral water', quoted in Henry Dale 1949: xix

30 'method of Sherlock Holmes': Marquardt 1949: 94

30 an intensely visual thinker: Marquardt 1949: 93

30 show up brightly: Heron 1882

30 In 1884: Gram 1962

31 In 1898: Lesch 2007: 18

32 a key in a lock: his Latin phrasing was *Corpora non agunt nisi fixata* – agents don't work unless they are bound. See Kaufmann 2008 and Marquardt 1949: 118

32 *Zauberkugel*: Bud 2007a: 14 and 225–226; Dutfield 2020: 153. The concept of the *Zauberkugel* from German folklore appears under the term *Freikugel* in a popular opera by Carl Maria von Weber, *Der Freischütz* (1821). The

first English translation of the term Ehrlich used was 'charmed bullets': Ehrlich 1908: 6

32 passed on in pregnancy: S. S. Wong 2016

32 around one in ten: Szreter 2014

32 'the great imitator': Hemarajata 2019

32 to produce a magic bullet: Ehrlich claimed to have already used dyes for treatment in 1891, when he used methylene blue to treat malaria in two patients. See Bosch & Rosich 2008

32 most organisms: as always, life finds a way. Marine sponges, which act as environmental accumulators, have evolved chemical means of dealing with high levels of arsenic e.g. Mancini et al. 2006

32 On 19 April 1910: Ward 1981

33 Ehrlich ordered: Marquardt 1949: 178

33 some people became almost violent: 'little short of violent'. See Ward 1981: 50

33 By September 1910: Ward 1981: 46

33 writing in pencil: Marquardt 1949: 180

33 water and caustic soda: Wechselmann & Wolbarst 1911: 73

33 fresh water: Marquardt 1949: 192

33 In October: Ward 1981: 51–52

33 In December 1910: Ward 1981: 45

33 arsenic that saves: Bosch & Rosich 2008

34 'the most remarkable': quoted in Ward 1981: 60

34 up to forty doses: J. H. Jones 1993: 46

34 55 per cent: Ward 1981: 50

34 by 1911: *Georg-Speyer-Haus Timeline*. Georg Speyer Haus. https://georg-speyer-haus.de/en/timeline/

34 6.6 million marks: as quoted in Cramer 2015, converted to modern equivalent using exchange rates from https://marcuse.faculty.history.ucsb.edu/projects/currency.htm (1914: $1 = 4.23 marks) and inflation calculator from https://www.usinflationcalculator.com/ ($1 in 1914 equivalent to $31.87 in 2024).

34 antisemitic caricature: a local conspiracy theorist and antisemite, Karl Wassmann, even filed a lawsuit against Ehrlich and his collaborator. Wassmann was later convicted of libel. See Zbar 2022: 69–70

35 concerned with patent laws: Dutfield 2020: 170

35 Under the German Patent Act: Dutfield 2020: 182

35 patent the manufacturing process: Dutfield 2020: 102

35 strike deals: Dutfield 2020: 180

35 'pirate nation' . . . 'country of counterfeiters': quoted in Fagin 2013

35 'never worked': quoted in Dutfield 2020: 182

36 price of a dose rocketed: Ward 1981

36 The *Deutschland* . . . July 1916: Koenig 1923: 277–278

36 Delivering Salvarsan by submarine: whether Salvarsan was specifically included as cargo on the first trip is unclear. The cargo was reported as 750 tonnes of 'medicinal and coal-tar dye products' of which dyes made up 125 tonnes, see Journal Of The Society Of Chemical Industry 1916: 1205. Others claim a larger consignment of dyes e.g. '400 tons' in Higgins 1919: 181. The *Deutschland* made a second trip in November 1916 to New London, Connecticut, when it also carried pharmaceuticals, see New York Times 1916. For more details and other contemporaneous articles, see Steen 2014.

36 complained in an open letter: Finney & Walker 1917

37 'either by accident or intention': New York Times 1917

37 In October: Ward 1981

37 dye industry was the foundation: Lesch 2007: 26

37 enemy-owned property: Trading With The Enemy Act 1917

37 Patricia died: Lesch 2007: 26 and Time 1929

38 began leasing: Lesch 2007: 26–27 and New York Times 1923

38 in the hands of the US: Speech from Mr. Treadway (Massachusetts) in US Congress 1919: 5738

38 Metz was furious: as quoted in New York Times 1923

38 'envy of the universities . . . wrapped in steam': quoted in Lesch 2007: 46

38 the singing of unfortunate canaries: Covell 1955; Linton 2010

38 Wilhelm Roehl: Holldorf 2003

39 'if one looks over': 1934 textbook, quoted in Lesch 2007: 38

39 unexpectedly died: Lesch 2007: 56

39 succumbed not to their wounds: Lax 2005: 14

39 Of the thirty-three: Lesch 2007: 53

40 dialogue with the medical researchers: see Lesch 2007: 63–65

40 all the mice treated: Lesch 2007: 61

40 'not all that different': quoted in Lesch 2007: 156

41 filing a patent on Christmas Day: Mietzsch & Klarer 1935

41 issues of timing and secrecy: Lesch 2007: 82 and 123

41 In early 1933: Lesch 2007: 90

41 by June 1934: see Lesch 2007: 82 and 123

41 unfortunately, French law: Lesch 2007: 123

41 Prontosil was successful: Lesch 2007: 104

41 preparing Christmas decorations: see Domagk's account as quoted in Lesch 2007: 105–106

42 they snapped the azo bridge: see Gingell, Bridges & Williams 1971 and Gingell & Bridges 1971

42 a 'pro-drug': see Fuller 1937 for contemporary discussion. For a recent review of drug metabolism by the human gut microbiome, see Zimmermann et al. 2019.

42 Just weeks before: Lesch 2007: 82

43 'completely useless': Lesch 2007: 126

43 In 1909: Gelmo 1908 and Lesch 2007: 82. The original patent was filed in 1909 as *Verfahren zur Darstellung von gelben Wollfarbstoffen* (Kaiserliches Patentamt Patent No. DE226239C).

44 the responsible person at Bayer: no inventors are listed on the original patent, but as the person in charge, Hörlein would have filed it. See Dodds 2017

44 like a gold rush: see Bickel 1988

44 entwined with the Nazi state: in 1947, 24 IG Farben executives including Hörlein were eventually put on trial at Nuremberg for war crimes and crimes against humanity. Hörlein was acquitted of all charges. See Lesch 2007: 109

44 *A Study in Scarlet*: quotes are from Doyle 1904

45 'mystique of colour': Lesch 2007: 84

45 emerged from fossil fuels: many drugs are still made from raw materials derived from petroleum. See Jimenez 2022

45 By 1940: Swain 1940

46 as early as 1907: Gradmann 2011

46 May & Baker: the drug was known as M&B 693, see Time 1944. M&B was a French-owned company based in Britain: Lesch 2007: 169

More lives than war can spend
Penicillin

47 'There is something to be learned about dust': *Irish Homestead* 1906, quoted in Bock 2007

47 an advanced microscope: Irving 1983: 68–69

47 Morell claimed: Trevor-Roper 1947: 60

47 bogus and impure: Irving 1983: 70–71. Wainwright 2004 writes that Morell may have used penicillin produced by the Allies and that he 'arguably did save Hitler's life'. In my opinion, both claims are far-fetched.

47 a teaspoon: in March 1940, Ernst Chain in Oxford had not much more than 40mg. See Bud 2007a: 29

48 a triumphant return: Bud 2007a: 60. A typical dose was around a million units; approximately 300 billion units were carried by Allied forces across the channel. See *D-Day: The Supplies*. U.S. Department of Defense. https://dod.defense.gov/Portals/1/features/2016/0516_dday/docs/d-day-fact-sheet-the-supplies.pdf

48 *Penicillium* can be found: Brefeld 1875

48 *P. roqueforti* and *P. camemberti*: Cheeseman et al. 2014 (yes, really)

48 *P. digitatum*: Onions 1966

48 *P. lacus-sarmientei*: Ramírez 1986

48 Bruce Chatwin: see Morrison 2010. Chatwin had recently contracted AIDS and his claims ('Trust me to pick up a disease never recorded among Europeans') should be taken as fantastical. *P. marneffei* remains a problematic pathogen for people with AIDS but it has been reassigned in the *Talaromyces* rather than *Penicillium*, making clear the difficulties of fungal classification. See J. F. Chan et al. 2016

48 'grassy tufts': Link 1809: 16–17

48 genetic analysis: Baldauf & Palmer 1993

49 One ancestor of *Penicillium*: discussed as 'one of the most debated conundra in the Palaeozoic record' by Honegger et al. 2017

49 branched off: Steenwyk et al. 2019. *Penicillium* emerged too late for *Diplodocus*, but possibly overlapped with *T. rex*.

49 50 million tonnes: Elbert et al. 2007

49 particulate matter: Womiloju et al. 2003 estimate 12–22 per cent (organic carbon) and 4–11 per cent (total mass) for fungi and pollen combined.

49 indoor fungi: Reboux et al. 2019

49 around five microns: a range of 2–5 μm is given by Fröhlich-Nowoisky et al. 2009

49 20,000 metres: D. W. Griffin 2004

49 In 1928: Fleming 1929

50 no trace: Laich, Fierro, & Martín 2002

50 an Irish scientist: Lax 2005: 25

50 *P. rubrum*: later, some mycologists subsequently classified Fleming's strain as *P. chrysogenum*. Recent analysis suggests that it was *P. rubens* and that La Touche's identification was accurate, consistent with taxonomic schemes at the time. See Houbraken, Frisvad & Samson 2011

50 *animal de laboratoire*: quoted in Houbraken, Frisvad & Samson 2011

50 much more plausible: Hare 1970: 81–87 argues that the laboratory window was shut and the *Penicillium* must have originated from La Touche's laboratory. Hare notes 'it is, of course, impossible to prove this theory'.

51 'like Stilton': quoted in Hare 1970: 90

51 clammy dressing: Lax 2005: 27–28

51 *Bacillus influenzae*: now known as *Haemophilus influenzae*. Fleming was preparing cultures as part of efforts to produce a vaccine for influenza, now known to be caused by the influenza virus rather than the (mis-named) *Haemophilus influenzae*.

51 mycotoxins: Perrone & Susca 2017

52 Jewish émigré: Bud 2007a: 29

52 Chain resembled: Lax 2005: 59–61

52 obtained years before: Lax 2005: 80

52 A unit measure: Robinson 1947: 87

53 biscuit tins: Lax 2005: 124

53 Heatley's device: Lax 2005: 141. When the Science Museum asked Heatley to build a replica of his machine in the 1980s, he was disappointed by how much it ended up costing, complaining that 'rubbish dumps aren't what they were in the 1940s.'

53 cobbled together one hundred milligrams: Lax 2005: 101

54 lurked in nearly all: Fleming reported 90 per cent for uniforms uncontaminated by blood or pus. See Lax 2005: 15

54 rubbed a pinch: Lax 2005: 126

55 'the key is man's power': Darwin 1859: 30

55 a series of valleys and hills: evolution is more complicated than this analogy suggests – see Levins & Lewontin 1985: 85–106. Lewontin stressed
that 'organisms do not adapt to their environments; they construct them
out of the bits and pieces of the external world' (quoted in Fromhage
& Houston 2022). For a discussion of the evolutionary importance of
understanding organisms as subjects rather than objects, see Godfrey-
Smith 2017.

56 at the edge of the cloud: here I am (deliberately) ignoring the important role
of phenotypic variability within genetically identical populations, referred to
as 'bet-hedging'. For a discussion, see Grimbergen et al. 2015.

56 by Christmas 1940: Lax 2005: 149

56 could destroy penicillin: Abraham & Chain 1940

57 animal tests: Bud 2007a: 30

57 Albert Alexander: see Sullivan 2022. Older accounts e.g. Bud 2007a: 32
and Lax 2005: 195 state that Alexander's infection came from a scratch
from a rose thorn while gardening – an incorrect claim that may have
arisen as wartime propaganda.

57 'oozing pus everywhere': quoted in Lax 2005: 195

57 'P-patrol': Bud 2007a: 32

57 died on 15 March: Lax 2005: 195

58 'Put your confidence': Churchill 1941

58 Just four days before: Lend-Lease Act 1941

58 Clipper seaplane: Lax 2005: 221

58 federal microbiologists: Bud 2007a: 34. 'mould kings': letter from Florey
on July 22 1941, quoted in Lax 2005: chap. 10

58 a tense few days: Lax 2005: 229

58 Marmite: Lax 2005: 127

59 straight into the Illinois: Scokeley 2014

59 sixty cents a gallon: Bud 2007a: 36

59 subject to official price controls: Sheehan 1982: 68

59 over fifty times more penicillin: Bud 2007a: 36

59 bodged extraction apparatus: Lax 2005: 249–250

59 Transport Command planes: Lax 2005: loc. 3407

59 cantaloupe melon: Bud 2007a: 36. Neushul 1993 quotes a mycologist who
employed Mary Hunt to search for strains saying this tale is 'folklore';

apparently a housewife brought it to the laboratory. This strain became known as NRRL1951. Successive mutations of that strain with UV, X-rays and nitrogen mustard (used for chemical warfare) led to a strain F.15.1 that yields 55 times the amount, see Hobby 1985: 234

60 Over a thousand chemists: Bud 2007: 50

60 Milislav Demerec: Bud 2007a: 38–39

61 Out of 5,000 mutated strains: Alfano 2016. Further mutagenesis led to more improvements in yield, see Backus, Stauffer, & Johnson 1946

61 A 2014 paper: Specht et al. 2014 studied the strain *P. chrysogenum* P2, a descendant of the Wisconsin 54-1255 strain produced during WWII.

61 scientists in other countries: see Bud 2007a: chap. 4

61 In occupied France: Bud 2007a: 79

61 In China: Bud 2007a: 80

61 In Russia: Bud 2007a: 81

61 In Germany, scientists believed: Florey was anxious to stop Fleming's strain falling into German hands. For Allied intelligence reports into German wartime production, see Shama & Reinarz 2002.

61 Mushroom Laboratories: Bud 2007a: 44

62 spare parts: Rodengen 1999: 65

62 over twenty companies: Bud 2007a: 45

62 'a race against death': *A Race against death* (LC-USZC4-1986) (1941). Library of Congress. https://www.loc.gov/pictures/item/91483071/

62 In North Africa . . . 'scallywags': details of this debate are from Howie 1979

62 98 per cent effective: Bud 2007a: 58. American military doctors would find that mixing penicillin with peanut oil and beeswax slowed down the drug's percolation; gonorrhoea could be cured with one (extremely painful) injection. See Romansky & Rittman 1944. The dreaded 'peanut butter shot' of benzylpenicillin ('bicillin') is still injected into the buttocks of new US Army recruits. See Kirkpatrick 2020

63 Judged by the price: Bud 2007a: 53. In 1943, the US government would pay $20 for a penicillin dose that would cost 6.5 cents by the end of the war. By 1953, this had fallen to less than a cent: 'the drug, only recently more precious than gold, was worth less than the bottles it came in'.

63 his Nobel lecture: Fleming 1945

64 Mary Barber: see Garrod 1966

64　Barber's own data: Barber 1947b.

64　tracking a resistant strain: Table III of Barber 1947b. Barber was adept at investigating hospital outbreaks, including tracing an outbreak of puerperal sepsis to the under-rim of a single lavatory seat, see Kenny & Barber 1944

64　Fleming himself proposed a method: Fleming 1942

65　As Barber showed: Barber 1947a. As she noted, 'an increase in penicillin-resistant staphylococci after penicillin treatment may not depend on acquired resistance, but may be the result of a process of selection – naturally sensitive bacteria being quickly destroyed and resistant organisms surviving'. She reiterated the point about the size of the inoculum in a later paper, see Barber & Rozwadowska-Dowzenko 1948

65　finish the course: for criticism of the messaging on this issue, see Llewelyn et al. 2017

66　'clever' bacteria that 'learn' to resist: for example, see the discussion 'I was mildly irritated to read a pamphlet in my doctor's waiting room . . .' in Dawkins 2009: chap. 5. In fairness to Fleming, newer research shows he was partially right: continuous low levels of an antibiotic can lead to the evolution of high-level resistance, but only if the bacteria are exposed for hundreds of generations. See Wistrand-Yuen et al. 2018

66　spread of penicillin continued: by December 1945, production was taking place in the US, UK, Australia and small amounts in Mexico. The US had begun exporting penicillin in June 1944; by this point, export equalled domestic consumption. Hobby 1985: 211

66　the waters at Lourdes: Sydney Morning Herald 1946

66　'Thanks to Fleming': Time 1955

66　recovery rate of 93 per cent: see Hastings 2010: 449. For a more sober assessment, see The War Office 1948: 281, which reports death rates from infective causes of less than 10 per cent even without penicillin. However, penicillin reduced these by between a factor of 1.7 and 13.9 for different categories of injury. For discussion of army studies of penicillin see Bud 2007a: 57

66　32 million yards of bandages: Hastings 2010: 448

67　One CIA report: Landas 2020: 140

67　The American state prevented: an exception was Japan. See Lee 2018

67　Smugglers profited: Landas 2020: 201

67 'a brisk trade': Landas 2020: 201

67 At a conference: Landas 2020: 184

67 A frustrated Ernest Chain: Landas 2020: 164

67 British complained bitterly: Bud 2007a: 69–71

68 'his penicillin': see Bud 2007a: 65

68 genetic analysis: Pathak et al. 2020

68 no legal owner: see Hobby 1985: chap. 6 note 1. There was no patent on the substance itself, though patents existed on manufacturing processes.

68 less than the bottle: Rodengen 1999: 74–75

Earthly powers
Streptomycin

71 'The soil is not': Waksman & Starkey 1931: vii

71 'a mere dot . . . godforsaken town': Waksman 1954: 17

71 'That odour': Waksman 1954: 38

71 chernozem: see Vysloužilová et al. 2016

71 Almost a quarter: Hasan 2022

72 *krotovinas*: see Pietsch 2013.

72 Waksman would walk: Waksman 1954: xi and 34

73 first course: Waksman 1954: 78

73 'more living things': Waksman 1954: 8

73 The bible on the topic: Chester 1901. The situation improved when the Society of American Bacteriologists updated the system of classification in 1917 and 1920, leading to the publication of Bergey 1923. Waksman is the first contributor thanked, for assistance with *Actinomyces* and *Thiobacillus*. For his first encounter with *Actinomyces*, see Waksman 1954: 81.

73 'to state': Buchanan 1916

73 accosted by a rabbi: Waksman 1954: 37

74 caught scarlet fever: Waksman 1954: 42

75 'I loved . . . so much': Waksman 1954: 29

75 he would state: Waksman 1954: xi

75 the author of one textbook: Tanner 1933: vi

75 'The insulted and repressed': Waksman 1954: 63

75 'How I Raised A Flock of Chickens': Waksman 1954: 72

76 René Dubos: details of his life are drawn from Moberg 2005

76 came down with a throat infection: Moberg 2005: 4

76 'the expropriation': see Marx 1890: 686 and 714

77 'robbing the soil': see Foster & Clark 2020

77 than Dubos found it: Moberg 2005: 7

77 In February 1923: Moberg 2005: 8

77 came across a piece: Moberg 2005: 10

77 'Soil is a living world': Winogradsky 1924 as quoted in translation in Moberg 1999 and Moberg 2005: 11

77 a Russian aristocrat: For biographical details of Winogradsky, see Dworkin & Gutnick 2012; Thornton 1953; Waksman 1946.

78 'characteristic functions': quoted in Dworkin & Gutnick 2012: 376

78 a few weeks after: Moberg 2005: 11

78 tall and imposing: Dubos, quoted in Moberg 2005: 12

78 'I know a winner': Moberg 2005: 12

78 a flash of insight: Greenwood 2008: 150

80 'uncouth earwax': Hotchkiss 1990: 3

80 one hundred thousand times: Honigsbaum 2016

80 5 trillion mice: Honigsbaum 2016

80 on his own garden: New York Newsday 1990

80 'like killing a tiger': New York Times 1939

81 'What has become': Waksman & Woodruff 1940

81 outlining protocols: Waksman 1945: 56–60

81 they were antibiotics: Jean Paul Vuillemin, who coined the term *antibiosis* in 1889, wrote that the situation described was 'so simple that one has never thought of giving it a name', see Brunel 1951. Waksman originally proposed 'antibiotic' as an adjective in 1942, see Bud 2007a: 107. Waksman himself claims he coined the term in 1941, see Waksman 1954: 207. From Waksman's perspective, gramicidin and penicillin were antibiotics because they were made by living organisms, whereas Prontosil and other synthetic chemicals were not.

81 over $250 a month: Pringle 2012: 27

82 Under the agreement: see Waksman 1954: 204 and Pringle 2012: 28 and 63

82 laboratory rats (actinomycin): Pringle 2012: 31, kidney failure (streptothricin): Pringle 2012: 33

82 Albert Schatz: for biographical details, see Auerbacher & Schatz 2006

82   Its ancestors had come from the soil: Gagneux 2018

83   killed more than 50,000: Pitney & Kasius 1947

83   less than 20 per cent effective: see Martinez et al. 2022

83   chicken's throat: Auerbacher & Schatz 2006: 35

83   called it streptomycin: Schatz, Bugle, & Waksman 1944

83   doctors at the prestigious Mayo Clinic: Rosen 2018: 197 In 1969, one researcher commented that, surprisingly, the original paper did not stress this discovery: 'No comment whatever on this sensational find is to be found in the text.' Quoted in J. F. Murray 2004.

84   clawed in desperation: Auerbacher & Schatz 2006: 33

84   By March 1944: Pringle 2012: 49

84   'deathly afraid': Auerbacher & Schatz 2006: 33

84   geosmin: Gerber & Lechevalier 1965. Waksman recalls how his professor at Rutgers told him around 1915 that he'd seen the actinomycetes many times before, but knew little about them, see Waksman 1954: 210

85   considered the possibility: Pringle 2012: 63

85   'absolute control': Waksman 1954: 204

85   he asked Merck: Pringle 2012: 63

85   an imposing presence: Hawthorne 2003: 24

85   Rutgers Foundation would handle: Waksman 1954: 204

85   Merck demanded repayment: Pringle 2012: 63

85   head of the government's War Research Service: Rosen 2018: 202

86   This argument had been used: Pringle 2012: 110

87   In September 1944: Pringle 2012: 65

87   'the fingers of my hand . . . any symphony otherwise': Waksman 1954: 203

88   Bugie signed an affidavit: Auerbacher & Schatz 2006: 47

88   the eventual patent: Waksman & Schatz 1948

88   November 1944: Pringle 2012: 57

88   receive a cheque: Auerbacher & Schatz 2006: 48

89   just three days before: Pringle 2012: 107

89   spontaneous generosity: Pringle 2012: 113

89   claimed to have sent out: Waksman 1954: 204

89   By 1947: Pringle 2012: 99

89   more than 99 per cent: Podolsky 2015: 19

89   over $187,000: Pringle 2012: 113, annual salary: Pringle 2012: 100

89   'partners . . . the discovery': Pringle 2012: 118

89 remarkable step: Pringle 2012: 137

89 beloved orchestra . . . twenty-four other: Pringle 2012: 144

90 in every detail: Wainwright 1991 was an important re-evaluation of Schatz's role as co-discoverer. For subsequent debate, see Kingston 2004 and Wainwright 2005.

90 'I suppose': as quoted in Bastian 2006

91 During his visit to Sweden: Waksman 1954: 308

Assault from all quarters
Isoniazid

93 'My experience in the sanatorium': quoted in Adair & Deuschle 1970: 90

93 Philip began to feel unwell: details from interview with Philip Shaw on 25 June 2021.

93 'shy, evasive': letter of 1924, as quoted in Sontag 1979: 7

94 established in 1932: *The History of Nunthorpe*. Nunthorpe Parish Council. https://nunthorpepc.org.uk/the-history-of-nunthorpe/

94 still powered by coal: see Mcneill 2001: 58–59

94 'a coughing': Mann 1927

95 'a dread disease': Dickens 1839

95 The newer antibiotic: for details of PAS, see Päuser 2012: 23–24

95 hopeless cases: Lehmann 1949

95 doctors around the world: Nagley 1950

95 By December 1949: see Medical Research Council 1950. The first randomised clinical trial conducted by the MRC had been on a dubious antibiotic called patulin. It had used a system of letters for each group of patients and had not been a total success: 'We wanted to muddle people up,' recalled the trial's designer. 'In fact we succeeded in muddling ourselves up. We didn't always remember what the letters stood for.' They learned from the experience to use random allocation using sealed envelopes when running the first trial on streptomycin. See Chalmers 2011.

95 when the antibiotics were used together: for laboratory experiments, see Graessle & Pietrowski 1949.

96 Philip's treatment: his recollections are not certain, but match the wider picture in England, where the combination of streptomycin and PAS was

being prescribed as early as 1950 and used routinely by the end of 1951. See Leeming-Latham 2015

96 Giorgio Baglivi: Daniel 2006

97 an amazing new medicine: for accounts of isoniazid, see Greenwood 2008: 170–179 and Päuser 2012.

97 Sea View: Ellis 2015; A. Griffin 2017

97 a quarter: out of over 20,00 beds in the city's hospital system, 5,008 were occupied by tuberculosis patients. New York Times 1953

97 press conference: Päuser 2012: 38

98 'I don't believe': quoted in Päuser 2012: 33.

98 'Due to Kogel indiscretion': quoted in Päuser 2012: 38

98 one part in 60 million: Päuser 2012: 30

99 on 17 December 1951: Päuser 2012: 30

99 'we should not worry about profits': Päuser 2012: 31

99 McDermott himself . . . had received: Beeson 1990

100 a plan for a staggered announcement: Päuser 2012: 38

100 When the Squibb CEO visited: Päuser 2012: 41, note 62

100 Lazarus-like: Kaempffert 1952

100 beyond anything they had ever seen: Kaempffert 1952

100 'KO for TB?': Washington Post 1952

100 'customary reserve': quoted in Päuser 2012: 40

100 One German magazine even claimed: Päuser 2012: 41

101 Robert Schnitzer: see Päuser 2012: 28

101 They agreed a joint statement: quoted in Päuser 2012: 41

101 over 150 grams: Päuser 2012: 26

101 'an inevitability': Päuser 2012: 41

102 'to a degree': quoted in Päuser 2012: 17

102 four-fifths of everyone: National Office of Vital Statistics 1953. Table 52 records that among 15–35-year-olds there were 8,858 deaths from infectious and parasitic diseases; of those, 7,039 were from tuberculosis.

102 Domagk wrote in December 1951: quoted in Päuser 2012: 36

102 the names of the centres censored: Irish Times 1952a

102 By a quirk of British law: Hillaby 1952

103 bamboozling variety: Irish Times 1952b

103 immediate fears: see Hillaby 1952 and Manchester Guardian 1952

103 isoniazid-resistant strains: Päuser 2012: 48

103 the shrine at Lourdes: Moberg 2005: 43

103 begun to cough up blood: Moberg 2005: 47

103 Four years later: Moberg 2005: 59–60

103 'wanted only': Dubos quoted in Moberg 2005: 60

103 as obsessed as him: Moberg 2005: 73

103 researcher on tuberculosis: Moberg 2005: 76. Like other scientists, she also contracted clinical tuberculosis, almost certainly from the dangerous experiments she carried out. Another, Bernard Davis, reported that researchers in the Dubos laboratory used unplugged mouth pipettes to move bacteria around – he once accidentally got a mouthful.

104 grand piano: Moberg 2005: 110

104 'exciting headlines': Dubos & Dubos 1987: 156

104 'the introduction of a new era': Cluff 1982

104 'a German dye': McDermott 1969

105 nine trips to hospital: Beeson 1990

105 from his bed: Beeson 1990: 286

105 the Navajo: for a historical overview of the Many Farms trials, see D. S. Jones 2002.

105 'regal poverty': McDermott, Deuschle, & Barnett 1972

106 spread of a respiratory disease: though tuberculosis was likely present on the American continent many thousands of years ago, its reintroduction by European colonisers was responsible for its devastating spread in Native Americans, exacerbated by a deterioration in living conditions after forced relocation to reservations. See Rieder 1989. More generally, the spread of tuberculosis lineages matches global events such as the Industrial Revolution. See Merker et al. 2015

106 flat and arid: for these geographical details see Adair & Deuschle 1970: 53

106 Infant mortality: McDermott, Deuschle, & Barnett 1972

106 at least the 1880s: Adair & Deuschle 1970: 9

106 a resolution: Adair & Deuschle 1970: 37

106 asked to see data: Adair & Deuschle 1970: 40

107 'We knew . . . right along': Adair & Deuschle 1970: 44

107 'It remains . . . like that': Adair & Deuschle 1970: 47

107 'We members': this quote and preceding details from Adair & Deuschle 1970: 54–57

107 'code talkers' in the US Army: Riseman 2007: 57

108 translation map: Adair & Deuschle 1970: 80

108 blood in their spit: Adair & Deuschle 1970: 125–126

108 'did not actually require': this quote and preceding details from McDermott, Deuschle & Barnett 1972

109 'almost to the bursting point': 1975 lecture by McDermott, quoted in Beeson 1990: 299

109 push their plane: Moberg 2005: 87

109 euphoria: Laurence 1952

110 45 per cent of them improved: Simpson 1954: 107

110 'the power behind the throne': Simpson 1954: 74

110 'he would no longer': Simpson 1954: 75

110 'theories': Simpson 1954: 115

110 'psychic energizer': New York Times 1957

110 $195 million a year: Health News Institute 1958

110 the 'chemical imbalance' theory: see Mukherjee 2022

111 around 400,000: New York Times 1958

111 'not a . . . drink': New York Times 1958

111 the antidepressant industry: see Lasker 1963: 405–406

111 'It seems safe to assume': Selikoff, Robitzek & Ornstein 1952

112 last tuberculosis patients left Sea View: Landmarks Preservation Commission 1985: 26

112 saplings took root: see photographs at Ellis 2015

Age of excess
Tetracycline

113 'It is assumed': McKeown 1979: 118

113 a seventeen-year-old: Rodengen 1999: 31

113 anxious vigil: Rodengen 1999: 63

113 Smith himself was dying: New York Times 1950

113 'Don't make . . . if we have to': quoted in Rodengen 1999: 80

113 global revenue of $100 billion: Dunleavy 2023

114 imported juice: see Rodengen 1999: 18–19

114 Mafia's formation: Dimico, Isopi, & Olsson 2017

114 in 1919: Rodengen 1999: 23, 30

114 *Aspergillus niger*: see Papagianni 2007

114 By 1924: Rodengen 1999: 33

114 an industrial secret: Rodengen 1999: 37

115 'if you want to lose': Podolsky 2015: 23

115 had opposed him: Rodengen 1999: 75

115 'chance favours': quoted in Pearce 1912

115 other new antibiotics: e.g. neomycin, Waksman & Lechevalier 1949

116 'This plodding': Fleming 1946

116 set a precedent: Costello 1968

116 seventeen-year period: this was extended to twenty years in 1995, see Kevles 2023

116 Parke-Davis sent: Rosen 2018: 244

116 Bristol-Myers Squibb sent: Maeder 1994: 75

116 Eli Lilly arranged: McGraw 1976: 80

116 Malayan soil: Nature 1956

117 135,000 . . . 20 million: Rodengen 1999: 76

117 'We got soil': Rodengen 1999: 77

117 'wanted a name': Rodengen 1999: 79. It seems probable that the nearby town of Terra Haute, Indiana, may have been the real reason.

117 died in July: New York Times 1950

118 Sackler's responsibility: Podolsky 2015: 25

118 'Terra bona': Radden Keefe 2021: 37

118 'one of the most complex': Podolsky 2015: 25

119 'broad spectrum of activity': quoted in Podolsky 2010: 331

119 are still taught: for criticism of the separation of antibiotics into 'broad' and 'narrow', see Acar 1997

120 initially mistook: Rodengen 1999: 79

120 piece of cardboard: Rosen 2018: 222

120 four-ring (tetra-cyclic) structure: Aureomycin had a chlorine atom so was called chlortetracycline ($C_{22}H_{23}ClN_2O_8$ – made of 56 atoms); Terramycin had an extra oxygen so was called oxytetracycline ($C_{22}H_{24}N_2O_9$ – 57 atoms).

120 \$1 of every \$4: Podolsky 2015: 28

121 less than a year: Podolsky 2015: 25

121 'a hundred wholesalers': Rosen 2018: 223

121 Fellow medical advertisers thought: Podolsky 2015: 25

121 'invented the wheel': quoted in Radden Keefe 2021: 38

121 killed over 450,000 Americans: Radden Keefe 2021: 4

121 purchased more than two-thirds: Rosen 2018: 229

122 librarians tore them out: Herzberg 2013

122 'is not the infant': Podolsky 2015: 145

122 Readers of *Good Housekeeping*: Kirchhelle 2020: 19

122 'on-location meat preservative': Landas 2020: 271

123 40 per cent of antibiotics sold: Podolsky 2015: 28

123 various brand names: Welch & Martí-Ibáñez 1960: 80

123 producing resistance to the others: Wright & Finland 1954

123 various filings and documents: for a detailed discussion of these arrangements, see Costello 1968.

124 'would have clearly been illegal': Costello 1968

124 'flash of creative genius': see Kingston 2001

124 help with the drafting: for Germany, see Dutfield 2020: 170.

124 'patentability shall not be denied' and 'non-obviousness': see Sampat 2015

125 federal economist John Blair: Rosen 2018: 276

125 fallen by two-thirds . . . prices stabilised: Costello 1968

125 profit margins of above 20 per cent: Podolsky 2015: 28

126 expenditure on drugs was triple: Bud 2007a: 111

126 companies responded angrily: Dutfield 2020: 184, 267; New York Times 1964

126 Finland analysed infection rates: Finland, Jones, & Barnes 1959

126 'total bacterial flora': Finland, Jones, & Barnes 1959. For more on the potential harms of antibiotics to the microbiome, see Blaser 2014 and Kinross 2023.

127 over sixty available: Podolsky 2015: 46

127 'rational therapy will give way to chaos': Podolsky 2015: 43

127 'to capture again': Podolsky 2015: 76

128 'One+One=Three': quoted in Landas 2020: 253

128 when a journalist attempted: Lear 1959

128 Pfizer bought 238,000 copies: Podolsky 2015: 81. Note that McFadyen 1979: 167 claims 'over 260,000'.

128 more than $280,000: New York Times 1960

128 raw-boned: Fontenay 1980: 3, owl-like: Fontenay 1980: 382

128 mob bosses: Fontenay 1980: 171

128 employed his long-time friend: Fontenay 1980: 112

129 The hearings highlighted questions: see Fontenay 1980: 380

129 their Pearl Harbor: 'Pearl Harbor Day', quoted in Tobbell 2011: 74

129 resigned the following day: McFadyen 1979: 168

129 Kefauver's committee concluded: Fontenay 1980: 383

129 promoted the 'drug story': Tobbell 2011: 70

129 as Kefauver pointed out: Tobbell 2011: 96

129 an attack on 'individual freedom': Fontenay 1980: 384

130 Kefauver set out a new vision: see Tobbell 2011: 80

130 a 'mere shadow': quoted in Fontenay 1980: 386

130 turned out to permanently stain: Shwachman & Schuster 1956; Wallman & Hilton 1962

130 even pure penicillin provoked reactions: by 1968, it was estimated that 0.015–0.04 per cent of people would have an anaphylactic reaction to penicillin treatment. Idsoe et al. 1968

131 around 1 in 40,000 cases: Baldry 1976: 151; Best 1967; Wallerstein et al. 1969

131 Doctors who had seen the side-effect: see Greenwood 2008: 223. Today, in countries with stricter pharmaceutical regulations chloramphenicol is no longer prescribed, outside of a handful of life-threatening infections, but in many countries it remains on sale for self-medication.

131 had only seven: Rosen 2018: 281

131 'an interesting collection': Rosen 2018: 282

131 given out over 2 million: Rosen 2018: 283

131 'Kefauver's fingerprints': Tobbell 2011: 118

131 the FDA's reviewers realised: Podolsky 2015: 3

132 Price-fixing lawsuits: for an overview of the complex set of class action lawsuits about antibiotic prices, see Wolfram, C. W. (1976). The Antibiotics Class Actions. *Cornell Law Faculty Publications*. Paper 1269. For the final opinion in the tetracycline litigation case against Pfizer see *United States v. Pfizer Inc.*, 676 F.2d 51 (3d Cir. 1982).

132 grew many times over: Knapp et al. 2010. The study is of Dutch soil samples, but is likely representative of global trends.

133 'Are we in medicine': quoted in Podolsky 2015: 152

133 rising to more than 4 million: Podolsky 2015: 87–89

133 William Kloepfer: Tobbell 2011: 9

133 would come to use similar techniques: for an overview of disinformation tactics see Conway & Oreskes 2012.

133 Kefauver bought fifty-one shares: Fontenay 1980: 392

Thesis, antithesis
Amoxicillin and clavulanic acid

135 'Nature designed': quoted in Laurence 1961

136 'decaying animal organic matter': Kadar 2019

136 Staphylococcal outbreaks: see Barber & Dutton 1958

136 In March 1956: Gillespie & Alder 1957

136 just one among many worldwide: see Jessen et al. 1969

136 from curtains to light fittings: Bud 2007a: 118

136 often contracted staphylococcal pneumonia: Bud 2007b

136 an American medical conference heard: Bud 2007b

136 'We may run out': quoted in Podolsky 2015: 146

137 in one London hospital: Barber et al. 1960

137 'This is London!': MacInnes 2011

137 upfront salary of $1 million: Humphries 2023: loc. 1059, Geist 1978: 308

137 Ancient Alexandria had been recreated: details from Kamp 1998

137 With his first glimpse: Geist 1978: 309

137 the reality of the British autumn: Humphries 2023: chap. 5

138 After two months of filming: Geist 1978: 311

138 Taylor should be replaced: New York Times 1961a

138 Taylor turned blue: Humphries 2023: loc. 987

138 Queen Elizabeth's personal physician: New York Times 1961b

138 an emergency tracheotomy: New York Times 1961c

138 brand-new antibiotic called methicillin: Laurence 1961. Taylor was also treated with phage therapy in the form of a staphylococcal bacteriophage lysate: New York Times 1961d

138 'out of danger': H. Thompson 1961

138 On 28 March: New York Times 1961e

138 She wore: Maitland 2022

139 she covered her face: these details from footage of the ceremony, see British Pathé 1961

139 He noted with amusement: Sheehan 1982: 14

139 over a thousand chemists: Bud 2007a: 50

140 synthesis efforts were largely abandoned: Sheehan 1982: 92

140 'disgustingly cheerful': Ferry 2019: 239

140 a fictional ship: Ferry 2019: 261

140 'You'd better take a gun': Ferry 2019: 263

140 one chemist had reportedly said: Ferry 2019: 246. The chemist in question, John Cornforth, claims he said he would take up croquet, not grow mushrooms (he did neither).

140 'diabolical concatenation': Sheehan 1982: 6

141 'anvil, hammer and tongs': Sheehan 1982: 7

141 he decided to use his newfound freedom: Sheehan 1982: 10

141 When else, he thought: Sheehan 1982: 16

141 He signed an agreement: Sheehan 1982: 128

141 variants of penicillin: for his suggestions for improvement, see Sheehan 1982: 132

141 by 1956, he had worked out: Sheehan 1982: 126

142 In 1957, he announced: Sheehan 1982: 157

142 'American lines': Corley 1994: 27

142 Beechams had enlisted: Corley 1994: 27

142 there were two ways of measuring: see Sheehan 1982: 170 and Bud 2007a: 125

143 down the drain: Sheehan 1982: 171

143 made an agreement with Beechams: Sheehan 1982: 160

143 at a fashionable London restaurant: Bud 2007a: 127

143 'personal communication': see BMJ 1959

143 'He is merely unskilled labor': Sheehan 1982: 176. Patent disputes about the new molecules did get nasty; Sheehan felt he should engage patent lawyers who were 'aggressive and competent' (Sheehan 1982: 175). Although Beecham and Sheehan had collaborated, their relationship descended into bitter litigation that rumbled on for twenty years. In February 1979, the US Board of Patent Appeals and Interferences finally found in favour of Sheehan. See Sheehan 1982: 197

143 first safety test: Bud 2007a: 128

144 on sale just eighteen months after: Greenwood 2008: 123

144 'theoretical objection': Elek & Fleming 1960

144 Barber had published a short letter: Barber 1960

144 'The proof of the pudding': Stewart 1960

144 Barber had found: Barber 1961

145 'eat some if not all': Stewart & Holt 1963

145 chemists were up to the challenge: 'How about an expression of faith in the adaptability of the chemist?' quoted in Bud 2007a: 122

145 'bird of ill omen': for these and other details, see Cornaglia 2000.

146 injected his mystery mould juice: Greenwood 2008: 115

146 Brotzu sent a copy: Greenwood 2008: 116

146 In 1959, they found that: Greenwood 2008: 118

146 'Cephalosporin Club': E. Jones 2001: 313

146 solved the problem in 1962: Morin et al. 1962

147 set ampicillin's initial price: Bud 2007a: 132

147 sales had increased by fifteen times: Corley 1994: 28

147 'stubborn versatility . . . go on forever': Asimov 1972: 650

148 came to be known as TEM: Datta & Kontomichalou 1965. Note that some accounts say that TEM was short for 'Temoniera' (τιμονιέρα) – supposedly the name of the original patient. However, this is unlikely: it is an old-fashioned Greek word for a steering wheel. Thanks to Aris Katzourakis for pointing this out.

148 taken by mouth: Greenwood 2008: 124–125

148 In February 1976: Percival et al. 1976

148 by 1981, it would be spreading: McCutchan, Adler & Berrie 1982

148 In May 1977, Beecham announced: Reading & Cole 1977

148 scientists at Eli Lilly: Higgens & Kastner 1971

149 Augmentin: Greenwood 2008: 132

150 arranged for the royalties from Japanese sales: The Waksman Foundation of Japan 2017

150 when Japanese microbiologists looked: Akiba et al. 1960

151 Watanabe and his colleagues: for a summary of this work, see Watanabe 1963

151 after being publicised: Bud 2007a: 176

152 If the cell is a computer: for this analogy, see Young 2016.

153 increase the rate of resistance in two ways: see San Millan et al. 2017.

153 the resistance genes didn't disappear: see e.g. Andersson & Hughes 2011; Lopatkin et al. 2017; Røder et al. 2021

154 as plasmids snowballed: Levy 2002: 106. For a genetic comparison of plasmids from before and after the introduction of antibiotics, see Cazares et
  al. 2024.

154 scientists found they could reconstruct: see Gillings 2014

154 lax cleaning practices: for contemporary complaints in the 1960s, see Bud
  2007b.

155 the most prescribed antibiotics in hospitals: Bush & Bradford 2016

155 'The ingenuity of nature': Greenwood 2008: 135

155 'When a victory is won': Sun Tzu 1993: 113

Last resort
Colistin

157 'American pigs and Soviet pigs': *Nikita Khrushchev's Visit to Iowa (part 2)*
  1959

157 'gloomy prophets': Podolsky 2015: 148

158 death certificates in England: Griffiths et al. 2004

158 Surveys showed that 1.5 percent of Americans: the figure of 17 per cent
  was from a New Orleans parish jail in 2007-2008, see McKenna 2010: 131

158 'other than AIDS': McKenna 2010: 77

158 an estimated 278,000 people: Klein, Smith, & Laxminarayan 2007

158 sixth leading cause of death: Klein, Smith, & Laxminarayan 2007

159 slowed dramatically in the 1980s: Achilladelis 1993

159 'me-too' drugs: see Greenwood 2008: 129 and Aronson & Green 2020

159 matched their order of first discovery: Clardy, Fischbach, & Currie 2009

159 they discovered daptomycin: McKenna 2010: 177 and 182

159 the soil of Mount Ararat: Eisenstein, Oleson, & Baltz 2010

160 between ten and fifteen years: Hall, McDonnell & O'Neill 2018: 61

160 from 1995, GSK spent seven years: Payne et al. 2007

160 more than a hundred reports: P. F. Chan, Holmes, & Payne 2004

160 a vulnerability window: see response quoted in House of Lords Select
  Committee on Science and Technology 1998: chap. 6

160 'all the signs are': Greenwood 2008: 136

160 scientists in New York reported: Bradford et al. 2004

161 CRE: Castanheira et al. 2008

161 around 2.6 million years ago: Pobiner 2013

161 Evidence from archaeology: for example, see J. C. Chen et al. 2024.

161 approximately doubled: Ritchie, Rosado, & Roser 2024b

161 over 40kg a year: Ritchie, Rosado, & Roser 2024b

161 on Christmas Day 1948: details of Jukes's experiment in what follows taken from McKenna 2017: 40ff.

162 vitamin B12 had been identified: Kirchhelle 2020: 36

162 B12 ultimately originates from bacteria: Martens et al. 2002

163 He made so much money: McKenna 2017: 42

164 'enormous long-range significance': quoted in McKenna 2017: 43

164 the FDA approved Aureomycin: McKenna 2017: 49

164 'universal lubricant': Kirchhelle 2018

164 'platoons of little pigs': Kirchhelle 2020: 20

164 'Feed AUREOMYCIN': image featured in Groeger 2012

164 by 1955: Kirchhelle 2018

164 Nikita Khrushchev: for biographical details, see Khrushchev & Khrushchev 2004

165 'I started working': speech given in *Nikita Khrushchev Goes to Hollywood* 1959

165 'the most advanced': Hill 2019: 14

165 threw silage at the massed journalists: Frese 2004: 152

165 farmer had himself visited: Brown 2013: 60

165 'little sausages' and 'corn is beefsteak': Hill 2019: 28

165 'There are many cases': this and other details in this section from *Nikita Khrushchev's Visit to Iowa* 1959

166 a popular Soviet joke: Hill 2019: 56

166 'petite . . . a super-pig' Smith 2010

166 'we can beat you in sausages': Frese 2004: 150

166 tending to form groups: Graves 1984; Landsberg & Denenberg 2014

167 Hungarian effort: Bud 1993: 32

167 'bone and fish meal': Polyanskiy Report—Pravda (23 December 1959) quoted in Kirchhelle 2018

167 'antibiotic proliferation': Kirchhelle 2018

167 by the mid-1960s: J. McGlone 2016; J. J. McGlone & Salak-Johnson 2009; Osborne Industries 2021

168 the pork industry was especially prone: Woods 2012

168 1947 had seen three publications: Greenwood 2008: 216–217

168 a company that manufactured soap: see footnote in Greenwood 2008: 217

168 destroying the kidneys: Hanif, Bali & Ramphul 2024 Falagas & Kasiakou 2006; Ordooei Javan, Shokouhi & Sahraei 2015

169 around 40 per cent of patients: Pogue et al. 2011

169 much better antibiotics were developed: Greenwood 2008: 219

169 from at least the 1980s: see bibliography of Savoini et al. 2002

169 brushed up against colistin: Bradford et al. 2004

169 'like my organs': Williams 2019

170 The rise of colistin resistance in Greece: Binsker, Käsbohrer & Hammerl 2022

170 over 300 times higher in animals: Binsker, Käsbohrer & Hammerl 2022

170 a complete ban: European Commission 2005

170 fifth-most common antibiotics: Catry et al. 2015

170 an estimated 12,000 tonnes: Davies & Walsh 2018

170 8 kg of pork a year: Brasch 2014

171 promoting large factory farms: K. Chen & Wang 2013; Gale, Marti, & Hu 2012

171 secret pork reserves: X. Wang 2020

171 half of all the world's pigs: Brasch 2014

171 new resistance genes: Walsh 2006

171 'the quiet before the storm': Walsh et al. 2005

172 Walsh was so shocked: these and other details from interview with Tim Walsh, 29 February 2024.

172 published by November: Y.-Y. Liu et al. 2016

172 Newspaper headlines proclaimed: Gallagher 2015

172 originated in bacteria called *Moraxella*: Kieffer, Nordmann, & Poirel 2017, respiratory tract of pigs: Larsen, Bille & Nielsen 1973

172 a random shuffling: Snesrud, McGann, & Chandler 2018

173 chickens in the 1980s: Z. Shen et al. 2016

173 heavy increase in colistin use: R. Wang et al. 2018

173 45 per cent of *E. coli*: C. Shen et al. 2020

173 nearly 50 per cent in November 2016: C. Shen et al. 2021

173 any number of possible routes: Y. Wang et al. 2017

173 The logbooks of freezers: European Medicines Agency 2016: tbl. 9

173 The global trade in live pigs: see e.g. G. G. R. Murray et al. 2023

173 wasn't only confined to pigs: for these details, see Supplementary Data 1 of R. Wang et al. 2018

173 a penguin: I am grateful to Fábio Parra Sellera for providing further details.

174 China was the most likely place: R. Wang et al. 2018

174 announced a ban: Chinese Ministry of Agriculture 2016

174 decreased from over 27,000 tonnes: Y. Wang et al. 2020 The decline in resistance may have been reduced by further evolution of the regulation of the gene that encodes MCR-1, see Ogunlana et al. 2023

174 Walsh worked with collaborators: Umair et al. 2023

174 96 per cent of colistin's global use: Y. Wang et al. 2020

174 they did not eliminate it: e.g. for China, see Li et al. 2024; Mai et al. 2023; Shafiq et al. 2023

175 high-rise blocks: Wakabayashi & Fu 2023

175 global forecasts: e.g. Mulchandani et al. 2023

175 not necessarily a key driver: for an example from England finding little overlap in bacteria between livestock and human infections, see Ludden et al. 2019.

176 MRSA from humans: McKenna 2010: 145

No viable path
Plazomicin

177 'Nature herself': interview with ex-USAMRIID scientist Hank Heine, 18 January 2024

177 formed from the compressed skeletons: Sloan 2006: 56–57

177 wealthiest city in America: Winters 2024

177 a wealthy industrialist had chosen: South San Francisco Historical Society 2020

178 a new company called Genentech: Mahrer 2023

178 founded by a biochemist: Genentech 2016

178 Harvesting one pound: Genentech 2016

178 looked like stuffed olives: Genentech 2016

178 DNA Way: see *Genentech: South San Francisco*. Genentech. https://www.gene.com/contact-us/visit-us/ssf

179 When Judice had started working: this and following details from interview with Kevin Judice, 7 December 2023.

180 an additional outer membrane: Tripathi & Sapra 2024

180 'If you want to make a new drug': as quoted by Judice. The usual form is 'The most fruitful basis for the discovery of a new drug is to start with an old drug'. See Raju 2000

180 Heinz Moser: interview with Heinz Moser conducted on 19 January 2024.

181 sisomicin: Weinstein et al. 1970

181 Looking out from his office: interview with Kevin Judice, 7 December 2023.

182 accident at a Soviet bioweapons facility: Gorvett 2017; Meselson et al. 1994

182 filled with dark red blood: New York Times 2001

182 Only eighteen cases: Stolberg 2001

182 nobody had died: L. M. Bush et al. 2001

182 terrified of opening their mail: Garrett 2011

182 a hazmat suit: Mayer 2008

182 cost of decontaminating buildings: Schmitt & Zacchia 2012

182 estimated at several billion dollars: L. M. Bush & Perez 2012; Heyman 2002

183 'outlaw regimes': The White House 2003

183 'less than a teaspoonful': Powell 2003

183 Iraq had no anthrax: before the invasion, Bush had cited 1999 UN weapons inspectors as saying that Saddam Hussein had the means to produce anthrax, but that weapons programme had been discontinued at the time Bush was speaking.

183 $5.6 billion: Project BioShield Act 2004

183 a self-confessed leftwinger: interview with Kevin Judice, 7 December 2023.

184 'I know what you've got': interview with John Hollway, 26 January 2024.

184 awarded the company $24.6 million: US Fed News 2006

184 ACHN-490: Endimiani et al. 2009

184 a further $18.8 million: US Fed News 2007

184 Phase 1 safety trial: Business Wire 2009a

184 $26.6 million contract: Business Wire 2009b

184 $64 million contract: Business Wire 2010

184 'flush with cash': Xconomy 2010

184 nearly $150 million: the vast majority was from the US government, but they'd also been awarded £4.1 million from the Wellcome Trust; in 2016, the Wellcome Trust owned 6 per cent of Achaogen. See The Legacy 2016

184 it even tried to return: Interview with John Hollway, 26 January 2024.

185 Kenneth Hillan: Business Wire 2011

185 the second-most advertised product: Frieden & Mangi 1990

185 only marginal activity: Donaldson, Palleti, & Carroll 1994

185 'Our experience': Frieden & Mangi 1990

185 fatality rate of one in two: L. Chen et al. 2021

186 the planned trial: Achaogen 2013

186 fewer than 3 such infections per 100,000: CDC 2013: 53

186 'extremely collaborative': Hillan 2014

186 yet another mark of distinction: GlobeNewswire 2015

186 acquisition target: Business Monitor 2015

186 Achaogen went public: MarketLine 2014

187 'huge profits': Money Morning 2014

187 countries with higher infection rates: Achaogen 2013

187 another Phase 3 trial: Sabih & Leslie 2024

188 published the Phase 3 results: GlobeNewswire 2016

188 more than doubled: 12 December 2016 saw Achaogen stock increase by +148.19%.

188 'Either you're having an affair': interview with John Hollway, 26 January 2024.

188 analysts raised their estimates: Business Monitor 2016; The Fly 2016a; 2016b

188 21st Century Cures Act: 21st Century Cures Act 2016

188 The Act would either: Kaplan 2016

188 'serious infection': 21st Century Cures Act 2016

189 over $1 billion: see *Achaogen Inc (AKAO)*. Barchart. https://www.barchart.com/stocks/quotes/AKAO

189 the highest possible status: WHO 2017

189 the FDA announced: GlobeNewswire 2018a

189 On the morning of 2 May: details in what follows are taken from the meeting transcript, see FDA 2018

189 the hotel was chilly: FDA 2018: 155

189 the numbers were so small: how the statistics were calculated would affect the conclusion. In the original trial protocol, Achaogen had planned to use one form of statistical test (a one-sided Z-test) but the final numbers were so small that they amended their statistical plan to use a different, less conservative method.

190 'non-inferiority' *was* supported: Achaogen's results suggested plazomicin was at most an absolute 5% worse than colistin, an accepted level.

190 might have been an impossible task: FDA 2018: 188

191 Shoshana Shendelman: FDA 2018: 221–226

191 'It's important to have': FDA 2018: 226

191 'We cannot lower': FDA 2018: 251

192 'An approval is an approval': FDA 2018: 344

192 Achaogen executives talked: FD Wire 2018

193 shares slumped by 28 per cent . . . 'yes and no': 24/7 Wall St. 2018; Benzinga 2018; Fierce Biotech 2018

193 continued to collapse: Watchlist News 2018

193 Medicare announced it would reimburse: GlobeNewswire 2018b

193 over $180,000: Beasley 2019

193 $56 per day: Wells, Nguyen, & Harbarth 2024

193 fortunes continued to worsen: see Wells, Nguyen, & Harbarth 2024

194 hadn't even made $1m: Carr, Stringer, & Shen 2020: 2

194 sat in a courtroom: interview with Ryan Cirz, 5 October 2024

194 'screwy commercialization dynamics': Lozos 2021

194 *E. coli* infections kill 36,000: Kevany 2023

195 beyond venture capital: see Hollway 2010

195 the first BARDA-funded antibiotic: see *Antimicrobials Program Portfolio Tracker*. CBRN Antimicrobials Medical Countermeasures Program. https://medicalcountermeasures.gov/barda/cbrn/antibacterials/

195 by 2024 amounting to $2 billion . . . over 130 different products: BARDA 2023

195 'no viable path': Farrar 2019

195 Before the end of the year: Gardner 2020; N. P. Taylor 2020; P. Taylor 2019

195 wouldn't even seek: Rex & Outterson 2020

196 stared down from his laptop: Cirz 2019

Realms of gold
Teixobactin

197 'The Lord hath created': King James Bible

197 According to one calculation: Hutchings, Truman, & Wilkinson 2019

198 perhaps less than 1 per cent: K. Lewis 2010

198 the Great Plate Count Anomaly: Staley & Konopka 1985

198 Researchers estimated from DNA: Rappé & Giovannoni 2003

198 archaea: for archaeal antibiotics, see Shand & Leyva 2008

199 brought the lab to the soil: Kaeberlein, Lewis, & Epstein 2002

199 iChip: Nichols et al. 2010

199 in 2015, they described: Ling et al. 2015

200 suggested that teixobactin challenged the dogma: Nature Podcast 7 January 2015, available at Ledford 2015

200 subsequent studies showed: Lloyd et al. 2020

201 'pretty chewy': Lowe 2015

201 challenging to synthesise: Jin et al. 2016; McCarthy 2019

201 the case of vancomycin: Outterson 2009

201 non-profit partnership called CARB-X: see CARB-X 2017

202 The money for CARB-X: see *Funding Partners*. CARB-X. https://carb-x.org/partners/funding-partners/

202 'This market failure': Orbinski 1999

202 used the prize money: DNDi 2023

202 In 2016: see *History*. GARDP. https://gardp.org/history/

203 GARDP would then own those rights: see GARDP 2023

203 'Nearly three years . . . take it over': Gallagher 2019

203 'a bit crazy': Gallagher 2019

203 pharmaceutical companies announced: AMR Action Fund 2020

203 a further $140 million: Eversana 2021

204 over $2 billion per drug: IFPMA 2024

204 structural challenges: see Wellcome Trust 2020

204 Silver has argued: Silver 2011

205 Outterson has estimated: Outterson 2023

205 teixobactin against anthrax: NovoBiotic 2020

205 still not started even Phase 1: NovoBiotic 2023

205 stirred hope: Grady 2015

205 'awful stability': Lowe 2015

206 tested over 48,000 molecules: Blasco et al. 2024

206 published an analysis of existing antibiotics: O'Shea & Moser 2008

206 Lipinski's rule of five: Lipinski et al. 2001

206 it had worked: Silver 2011

207 started mildly enough: this and further details from LaMotte 2022

207 'Iraqibacter': Scott et al. 2004

207 Wars had a track record: see Abdurasulov 2025 and O'Neill 2024

207 up to 35 per cent: Antunes, Visca, & Towner 2014

208 'if you can hear me': LaMotte 2022

208 The book told the life story: Lewis 1925

208 phage therapy became outmoded: it continued post-WWII in the Soviet Union, but also in other countries including France. For a recent collection of articles revisiting the history of phage therapy, see Sankaran 2020.

209 'literally had speed dial': quoted in LaMotte 2022

209 She believed her husband: LaMotte 2022

209 published about the results in 2017: Schooley et al. 2017

209 working around the clock: LaMotte 2022

210 In a review: Strathdee et al. 2023

210 The number of new clinical trials: Strathdee et al. 2023

210 Martha Clokie acknowledged: University of Leicester 2023

One wide expanse
Halicin

211 'One should sit quietly': The Paris Review 1990

211 at least 10^{60} possible molecules: Ball 2015; Reymond 2015

211 imagined by Jose Luis Borges: Borges 2001: 216

211 than there are stars in our universe: the number of stars is generally estimated to be around 10^{24}. See e.g. Traversa-Tejero 2021

212 without life, half of all minerals: Hazen & Morrison 2022; J. Thompson 2022

212 isoniazid is a derivative: Loots, van der Westhuizen, & Botes 2007

212 'All of the "easy"': quoted in Hall, McDonnell, & O'Neill 2018: 58

212 donated a state-of-the-art computer: for this and following details, see Martin 2018.

212 cost nearly $200,000: $18,000 in 1961 converted to 2024 dollars, see Weik 1961: 222

213 on one model: Martin 2018

213 they discovered a new law: Hansch et al. 1962

213 In 2020, researchers at MIT: Stokes et al. 2020

213 thought at first it was a tedious project: interview with Jon Stokes, 15 February 2024.

213 Drug Repurposing Hub: Corsello et al. 2017

214 over 7,000 molecules: *Drug Repurposing Hub*. Broad Institute. https:// repo-hub.broadinstitute.org/repurposing

214 'An original idea': Fry 1991: 39

214 Newspaper headlines dutifully proclaimed: Sample 2020; Trafton 2020

215 Phare Bio's stated aim: Collins Lab 2020

215 a second antibiotic candidate: G. Liu et al. 2023

215 abaucin: I wrote about abaucin and halicin at the time, see Shaw 2023a

215 recovery after a heart attack: Jang & Javadov 2014

216 'just another toxic chemical': Rex 2020

216 They hoped to take two years off: Floersh 2023

216 only progressed to animal studies: email from Akhila Kosaraju, 8 February 2024.

216 $2.6 billion per drug: DiMasi, Grabowski, & Hansen 2016

216 'if you believe that': MSF 2014

216 as low as $161 million: Schlander et al. 2021

217 'Health care is big business': Deloitte 2019

217 launched first in the US: Mulcahy 2024

217 Ryan Cirz has estimated: Cirz 2021

218 still tell stories: email from Alexander Bieri, 15 January 2024.

218 Ciba-Geigy merged: Grimond 1996

218 Novartis announced it was ceasing: Kasumov 2018

218 only a quarter of the companies: Lo 2015

218 Twenty-seven pharmaceutical companies: Rapp 2004

218 'bad for science': quoted in Lo 2015

218 researchers analysed patent filings: Haucap, Rasch, & Stiebale 2019

218 heavy reliance on costly research and development: compared to other sectors of the economy, pharmaceutical companies spend a lot on R&D. For a discussion, see Lowe 2013.

218 ten times as much on research: Congressional Budget Office 2021

219 estimated to be only 3,000: AMR Industry Alliance 2024

219 twenty times more patents awarded: AMR Industry Alliance 2024: 4

219 as much of 80 per cent of antibiotic development: as quoted by Gitzinger 2024, based on data from WHO 2022

219 ten employees or fewer: Gitzinger 2022

219 less than €1 million: Gitzinger 2024

219 only 10 per cent remain in the field: AMR Industry Alliance 2024: 14

220 'completely stupid': interview with Heinz Moser, 19 January 2024

220 the fire extinguishers of medicine: Rex 2024

220 based on provision rather than use: for a review of antibiotic resistance from an economic perspective, see McDonnell et al. 2024.

221 In 2022, the British government: Mahase 2020

221 signed a contract with NHS England: NHS England 2022

221 proportionate to its population: NICE 2022: 3

221 willing to pay up to £20,000: NICE 2012

221 too early to judge: Glover et al. 2023

221 awarded the £8 million prize: Challenge Works 2024

222 the prize of a voucher: Boyer, Kroetsch, & Ridley 2022

222 now supports it for the same reason: interview with Kevin Outterson, 6 March 2024.

222 'orphan drugs': see Mikami 2019

223 a 2021 study: Chua, Kimmel & Conti 2021

223 over $4 billion in sales: Becker 2024

223 The desire to avoid direct payments: for a discussion, see Hall, McDonnell, & O'Neill 2018: 93

223 The risk is that: for a review of commitments and actions on antibiotic resistance with economic modelling, see Global Leaders Group on AMR 2024.

223 announced in June 2024: Eli Lilly 2024

223 At the end of 2023, F. Wong et al. 2024

224 Stokes believes: Floersh 2023

225 Lee had predicted: Zastrow 2016
225 'Here?!': Hui 2016
225 in 2025, a new study: Goldberg et al. 2025
226 'our ingenuity': quoted in Wainwright 1990: 172

Conclusion:
the art of healing

227 'A conquest of this kind': de Beauvoir 1972: 157
227 by far the wealthiest: Federal Statistical Office, Switzerland 2023
227 'English understatement': quoted in Dornberg 1994
227 Paracelsus: see Porter 1998: 201–205
228 shoe-buckles contained more wisdom: Parker 1915: 732
228 'If disease puts us to the test': Pachter 1951: 57
228 'The art of healing': Paracelsus 1988: 50
228 'chemical medicine': Porter 1998: 209
228 Goa stones: Bailly 2014; Breen 2024
230 'no reason why anyone': Welch & Martí-Ibáñez 1960: 132
230 the majority of staphylococci worldwide: Bush & Bradford 2020
230 reduce transmission by 97 per cent: Moseley et al. 2024
230 second-leading cause of preventable stillbirth: see *Mother-to-child transmission of syphilis*. WHO. https://www.who.int/teams/global-hiv-hepatitis-and-stis-programmes/stis/prevention/mother-to-child-transmission-of-syphilis
230 an Al Jazeera report: Guimaraes 2017
230 the only remaining penicillin factory: Macho 2023
231 over fifty times: see Figure 5 of Hunermund, Schmidt-Dengler & Takahashi 2014
231 Pfizer returned a net profit: see *Pfizer Profit Margin 2010-2024*. Macrotrends. https://www.macrotrends.net/stocks/charts/PFE/pfizer/profit-margins
231 by 2023 it had fallen: see *Tuberculosis (TB)*. WHO. https://www.who.int/news-room/fact-sheets/detail/tuberculosis
231 in 2020, tuberculosis research received: Tomlinson 2021

231 Phumeza Tisile: details are drawn from Tisile 2014 and Tisile's blogs for MSF during her treatment.

232 lying alone in a dark room: Tisile 2021a

232 which made her deaf: Stoker 2023: 9

232 visit a priest: Segar 2023

232 'If I could change anything': Tisile 2021b

232 Bedaquiline: for the patent, see Van Gestel et al. 2003. For FDA approval, see Walker & Tadena 2012. I wrote about bedaquiline in Shaw 2023b.

232 hailed as a game-changer: see e.g. Brust 2023

232 by filing secondary patents: for secondary patent, see Hegyi et al. 2007.

233 only a few thousand were receiving: Thiagarajan 2023

233 In 2019, the Indian government: BMJ 2019

233 filed a legal challenge: MSF 2023

233 J&J then announced a deal: Gordon 2023

233 'creative procurement solution': Treatment Action Group 2023

233 In September: Malan 2023

234 J&J eventually agreed: MSF 2024

234 the public money: specifically, between 1.6 and 5.1 times J&J's own investment, see Gotham et al. 2020

234 'twelve months of Covid-19': Stop TB Partnership 2021

234 in early 2025, the US government: Stop TB Partnership 2025

234 'It never made sense': Stoker 2023: 13

235 Since 1990: Naghavi et al. 2024

235 The OECD estimates: these and other statistics in this paragraph taken from OECD 2023 unless otherwise cited.

235 grew by more than 11 per cent a year: Vihta et al. 2018

235 life expectancy will fall by around two years: Global Leaders Group on AMR 2024

236 She has taken phonecalls: Brenda Waning, AMR Conference Basel, 7 March 2024.

236 it was believed to be unethical: see Gabriel 2020

236 'would probably not': Notegen 2012: 11

237 'not to do things': Keynes 1931.

237 As one group of academics suggested: Singer, Kirchhelle & Roberts 2020

237 would pay for itself: for estimates of a return on investment of at least 7x, see Global Leaders Group on AMR 2024.

238 an atmosphere of paranoia: see Eisen 2017

238 some researchers have even advocated: Klug et al. 2021

239 grand technological endeavours: for figures see Singer, Kirchhelle, & Roberts 2020

239 As drug developer John Rex admits: 'capitalism works great for everything except antibiotics' Rex 2024

239 antibiotics are part of the infrastructure: Chandler 2019

240 when I put the suggestion to: interview with Kevin Outterson, 6 March 2024

240 'I've got something': quoted in Bud 2007a: 27

Index

isoniazid 93–112, 139, 212, 232, 236
  depression and 110–11, 112
  discovery of 97–101
  euphoria side-effect 109–10
  new era in treatment of
    tuberculosis and 111–12
  patents 100–102
  resistance and 95–6, 103, 104, 111
  trials 105–10

Japan 149–51, 157, 168, 173, 174
Johnson & Johnson (J&J) 223, 232–4
Judice, Kevin 179–85
Jukes, Thomas 162–3

Kafka, Franz 93, 234
Kefauver, Estes 128–32, 133
Kefauver-Harris Act (1962) 128–32, 133
Kelsey, Frances 131
Kennedy, John F. 130
Keynes, John Maynard 237
Khrushchev, Nikita 157, 164–7, 170–71
Kirchhelle, Claas 6, 167
Kloepfer, William 133
Klondike, Canada 1–2, 5
Koch, Robert 21–3, 83, 108, 198
Kogel, Marcus 97–100, 102–103
Kolkata, India 21

La Touche, C. J. 50
Labour government (1945–51) 126
Lamarck, Jean-Baptiste 13–14, 21
LaMattina, John 218
Lane, Nick 17
Latour, Bruno 20, 23
Lawal, Mashood 8
Lederberg, Joshua 6
Lederle 119–20, 123–4, 125, 162–4
Lee Sedol 225

Leeuwenhoek, Antoni van 12–13, 19, 20, 73
Lehmann, Jörgen 95
Lend-Lease Act (1941) 58
Levy, Stuart 16, 247
Lewis, Kim 199, 200
Lewis, Sinclair 208
Linnaeus, Carl 13
lipids 168–9, 172, 204
Lipinski's rule of five 206
London International Exhibition (1862) 29–30
Lowe, Derek 201, 205

magic bullet 32–4, 42, 44, 45, 110, 204, 207, 210, 215, 240
malaria 2, 27–8, 38–9
Manhattan Project 67
Many Farms 105–9, 112
Margulis, Lynn 11
Marx, Karl 76–7
mauveine 29
May & Baker 46
Mayo Clinic 83–4
McDermott, Walsh 99, 104–109
McKeen, John 113–15, 117, 120, 132
me-too drugs 159
meat consumption 161–3
Médecins Sans Frontières (MSF) 202, 216
Medical Research Council 95
Medicare 193
meningitis 87
Merck 81–2, 85–8, 115, 139, 141, 162–3
Merck, George Wilhelm 85–6
mergers, corporate 217–18
methicillin 138–9, 143–5, 158
Metz, Herman A. 36–8
microbiology 67, 75, 76, 77, 80, 82, 150, 198, 199
microbiomes 18, 42, 64, 127, 151, 153, 210